CONTENTS

		Page
1	Materials in Ancient and Modern Civilization	1
2	New Awareness	14
3	Traditional Practices	21
4	Knowledge, Training and Standards	37
5	The Necessity for Design	46
6	Materials and Ideas	54
7	New Factors Affecting Design	68
8	Design Education	78
9	Materials and Meaning	92
10	Providing for Primary Education	100
11	The Necessary Range of Materials	110
12	Further Steps for Education	122
13	Conclusion	130
	Bibliography	132
	Index	134

'The tool precedes intelligence.'
Andre Leroi-Gourhan

18. FEB. 98

m9400935
12.95
MABA

BI 0784175 2

This item may be recalled before the date stamped below.
You must return it when it is recalled or you will be fined.

NORMAL LOAN

KENRICK LIBRARY
PERRY BARR

Published in the United Kingdom by
The Design Council
28 Haymarket
London SW1Y 4SU

Printed and bound in the United Kingdom by The Bath Press, Bath
Designer: Nicole Griffin
Assistant designer: Carol Briggs

All rights reserved. No part of this publication may be reproduced, stored in a retrieval system or transmitted in any form or by any means electronic, mechanical, photocopying, recording or otherwise, now known or hereafter invented, without prior permission of the Design Council.

© John Fulton 1992

British Library Cataloguing in Publication Data

Fulton, John
    Materials in Design and Technology.
    I. Title
    620.11

ISBN 0 85072 289 6

Picture acknowledgements: Norman Manners p22, p89, p113, p126; Deanne Dewer p108; The Design Council p76.

PREFACE

A large proportion of the present generation, in what is often called the consumer society, is more familiar with manufactured goods than it is with the materials from which they are made and the processes employed in forming them.

I welcome the advent of a curriculum for design and technology as an indication of a national resolve to raise a new generation which understands technology and is capable of designing its responsible use. I have attempted to show how materials, and the technologies they gave rise to, have been important in the development of human intelligence and continue to be important in the education of children.

The book points to the intimate relationships which exist between materials, technology, language and culture, and argues that access to the knowledge and concepts of our time may depend more than we have realized upon familiarity with the materials upon which they are modelled. If this perception is sound it has implications for a rational approach to the provision of a suitable range of experiences with materials, particularly during primary education.

Some teachers have found these perceptions useful in clarifying their approach to their own provision of materials and have, therefore, encouraged the writing of this book. I am conscious that although these perceptions have survived my own extensive experience as a teacher, they remain limited and therefore need to be tested upon the experience and expertise of others.

Although the views expressed in this book are my own, I would like to thank all those colleagues, teachers, pupils and students whose example prompted the writing of it. I would also like to thank Margaret, Jane, Clare, Paul, Oliver and Rachel, each of whom has helped me over a number of passages of slow learning.

In particular I wish to thank Ivan Davies, for his patient and perceptive reading of the manuscript and for his helpful suggestions, and also the staff of the Design Council for their editorial guidance.

John Fulton
November 1991

THE AUTHOR

John Fulton's 26 years of teaching experience include periods at a junior technical school and school of art, as well as in teacher training. He has held lectureships at both Matlock and Northumberland Colleges of Higher Education.

John Fulton has worked in course design and development for professional bodies such as the CNAA Art and Design Education Panel and the DES. He was appointed HMI with responsibility for developments in design and primary education.

Since 1987, he has returned to teaching and acts as a consultant in art, craft and design education.

1 MATERIALS IN ANCIENT AND MODERN CIVILIZATION

FAR from being something new in the life of homo sapiens, technology has been present since before the birth of the species. The fact that humankind has grown up with materials and technology must be considered important in any account of their educational value. Discussion will not, therefore, be confined to what occurs, or might occur, in schools but is intended to consider the wider context in which activity with materials and experience of technology contribute to human capability and learning. Its purpose is to assist teachers as they develop their own perceptions for the appraisal of their practice and evaluation of the responses of children.

The discussion will emphasize the importance of physical matter in shaping both body and mind, and the importance of technology in shaping ways of thinking. It seeks to offer, as a necessary rationale for enterprising teaching in design and technology, a more thorough understanding of materials and what can be learnt from them.

It is useful to look at the origins of humankind and ask, 'What is it about human beings that has led to our using materials and developing technology so much more extensively than other species?' First, it is necessary to clarify what is meant by some of the words in the question before looking for a useful answer to it.

The word 'materials' here means any physical matter from which things can be made or which can be used for a specific purpose. It therefore includes all substances in whatever forms are found to be useful. The word 'technology' refers to all the means and processes by which the useful potential of all materials is realized, and the most important of these processes is the use of one or more purposefully shaped materials, in

INTRODUCING MATERIALS

the form of some kind of tool, to work upon another material.

The origins of technology, along with the origins of language, are lost in the extended period of human and pre-human life before the existence of written records. However, archaeological evidence reveals that tools of considerable refinement existed long before the arrival of homo sapiens (the human species). Knowing this can help us understand how the readiness of some of our ancestors to put materials to use prepared the way for the development of human intelligence.

The higher primates known as hominids, who existed four million years ago, displayed characteristics which allowed for technological development. Among the ways hominids differed from other species was in the genuinely upright carriage of the body, which allowed full use of shoulders, elbows, hands and fingers. Two-legged mobility, useful arms and hands, and forward-looking eyes and stereoscopic vision were inherited from forest-dwelling ancestors but were combined, in more open and varied terrain, with greater versatility and carrying capacity. These characteristics enabled some hominids to explore a wider environment and respond, more variably than other species, to what they encountered.

Amongst their more persistent encounters were those with the surrounding materials, so the next question should be, 'What was it about materials that generated the technological and mental activity which contributed so much to human inheritance?'

Valuing the limitations of materials
Today we are familiar with the fact that new materials enable us to make new products and develop alternative technologies. However, in order to understand the long-standing significance of materials in our lives we need to note that throughout the longest period of human development, the availability of materials was limited to those which could be used much as they were found. Whether easily accessible as objects, like sticks and stones, or discovered as earthy substances, like chalk or clay, each material came with its particular range of limitations or possibilities which were influential upon the lifestyles of their users.

The structure of what can be made in any material can vary but only within limitations imposed by the nature of the material itself and the tools that are able to shape it. For example, natural fibres could be spun and knotted to form a net which could be used for a variety of purposes, like an animal trap or a hammock. Either object makes positive use of the limitations of twisted fibres. Other traps or beds could be made from reeds or saplings but their form would need to be quite different to take advantage of the limitations imposed by the various materials. This continuous task of turning structural limitations to advantage and realizing their possibilities has been the main way in which materials have provoked technological and mental activity.

Most materials are still used in ways which reflect the earliest discoveries made about them. They are often used alongside modern equivalents: reeds for thatching are used with fibre-glass insulating material, showing that not only the materials but also their principles survive in contemporary technology.

Leroi-Gourhan *(Daumas 1962, p16)* observed that technology is cumulative and demonstrated how all known forms of pottery production, beginning with the most primitive, can still be found in some parts of the modern world, and that hard physical labour coexists alongside sophisticated applications of atomic energy.

The development of human ways of living were bound up with the materials that our ancestors were able to utilize. These included those materials which fell into the ancient categories of earth, fire, air and water and also from animal and vegetable sources, including the material structure of their own bodies and the physical conditions of their environment. Where some stones could be heaped together, others could be scratched or chipped; clay or dung could be pressed into new shapes; gut and fibres could string or tie things together. It is to such useful properties that the human species owes the origins of tools and skills, together with the mental abilities which accompany them. These latter abilities also enabled us, through models, to build upon, mark, shape and put together *ideas* as well as *materials*.

This two-year-old has already acquired sufficient control to sit upright within the confines of a metal bowl on sloping ground. He holds an aluminium jug and a polythene mug. To prevent the mug's rotation under the weight of added water he has extended his little finger; he is already familiar with the materials and forces involved.

Some challenges remain as they were in primitive societies; for example, one physical attribute, the ability to fetch and carry, contributes a challenge to the mind and can lead to continued learning. Eventually, the mere fact of being able to carry things from one place to another imposes the need to make choices. It leads to improvising containers for carrying and storage and then to bartering. A carried object may become a tool, a weapon, a token or a sign, as circumstances require. Each such development leads to social and environmental change and requires the successful adaptation of individuals and groups.

The more our ancestors could do the more they needed to learn what to do. In the absence of formal tuition, learning what to do meant learning to recognize from experience, sometimes by trial and error, what was the most appropriate action for a particular circumstance. Enhanced powers of exploration extended the environment and varied still further the conditions under which behaviour had to be adjusted. These factors called for increased discrimination: the ability to see possibilities and to choose between them. Individuals or groups which could not meet that demand

PHYSICAL ATTRIBUTES AND THE MIND

would stand much less chance of survival than those who could, and this circumstance favoured those with enhanced perception and a capacity to learn. This is the first indication that learning to make wise choices between perceived alternatives may be related to survival.

Extending physical attributes
If biologically inherited characteristics led to the use of tools, the tools themselves dramatically developed those characteristics. No longer limited to what could be achieved by hand, there was potential for an entirely new relationship between the human species and the environment. Every extension of the limbs, which tools provided, also extended the mind and the senses. In this way, technology enhanced the very characteristics that had led to its development.

J Z Young observes that:

> 'Perhaps the most important of all these characteristics of primate life for the development of man was the tendency to investigate and observe, and then to remember the results and so plan each individual life, and if necessary to seek out new and diverse environments. These activities were certainly stimulated by tree life and the consequent development of senses and brain. They led, together, to the appearance of the phenomenon we call mentality or thinking, dependent upon the use of internal models in the brain.' *(1974, p472)*

Throughout many millennia the 'higher' primates, and later homo sapiens, learnt to exploit the capacities of their own bodies and the materials around them. Control of their bodies was learnt through contact with earth, rock and vegetation on a large, environmental scale, while control on a smaller scale involved handling, smelling, tasting, looking at and selecting smaller quantities of materials with different qualities, properties and potential.

Early tools and weapons
Because pebbles and stones, particularly flints, are not easily destroyed they provide the oldest surviving evidence of how materials were used: it appears that stones have been modified for use as tools and weapons for two million years. Their evolution illustrates the development of technology from pre-human times to the late Stone Age. By examining flint tools and the sites where they were made and used, anthropologists have been able to deduce how they were produced and were also able to form some impressions of the 'societies' to whom they belonged. Each refinement from axes to scrapers and from knives to spearheads might represent thousands of years and profound changes in lifestyle.

Materials in ancient and modern civilization

Fire-hardened spears, and the prevalence of tools for scraping hides, support the view of some anthropologists that Neanderthals, who lived from 200,000 to 30,000 BC, killed, cut up and cooked large animal prey before the development of arrows, spearheads or bows. They also buried their dead and appear to have cared for their old. These features suggest a degree of social development, teamwork and the communication of hunting strategies, including trapping, which might imply other skills such as tying, lashing and sharpening timber staves.

It can be assumed that for each of the long stages of development for which there is tangible evidence, there would be comparable development in technologies based upon other materials. This is a reasonable assumption since all other raw materials would not only disappear more rapidly than flint but would offer less resistance to being worked in the first place. Flints were certainly not the first tools to be used nor the first material to respond to manipulation. The earliest hand-made flint scrapers indicate a persistent intention and skill which could only have been acquired from earlier success in handling easier materials with less permanent tools. For example, the scraping of animal skins with flat bones, like shoulder blades, was certainly possible but would have led to recognizing the need for a more efficient tool. The use of a broken antler as an adze (a tool with an arched blade at right angles to the handle), or long bone as a weapon would reveal limitations which could be overcome by making the implements heavier with the addition of a stone.

Was the first stone attached to a club to lend weight to it or was a shaft added to a stone to make it easier to handle effectively? Whatever the initial reason, it does not follow that the alternative advantage would be perceived immediately. The earliest form of stone tool remained virtually unchanged for a million years, after which tools were developed to become be more varied and adaptable.

Early human existence
It is necessary to bear these timescales in mind when considering the development of human intelligence. There were to be almost another million years of tool-making before, towards the end of the last Ice Age, recognizable human life emerged with husbandry and agriculture, settlement and town life and the kinds of social organization that make such life possible.

Jelinek writes:

'The period in which man, during his development, has lived as a hunter and collector was considerably longer than any of the other following developmental stages in which he started farming and animal breeding, and began to form tribal societies. The study of the earliest periods of man's existence gives us an insight into the basis of human development and the material and intellectual roots from which it all sprang, as well as being able to explain and understand the oldest and most important discoveries of man.' *(1975, p592)*

For all but the last ten thousand years, human beings lived in small groups gathering, hunting and fishing, avoiding predators and adapting directly to the conditions and material resources which they encountered. It is in this part of our past that we must locate the origins and nature both of technology and 'thinking'.

The importance of understanding the origins of technology and the characteristics of materials is that we may begin to see that they have not only enabled us to do things but have provided us with the provocation and the means to think in ways which were not previously possible.

> 'The immense importance of technology in moulding how we think has implications still only dimly appreciated …. It does however imply that any psychology based only on our biological origins is going to be inadequate: what is amazing about man is the extent to which he escaped his origin. This is through the use of tools, and the effect on us of technology that the tools have created. These considerations are at least as important as our biological background for understanding man and how the first mythological, philosophical and scientific ideas were conceived and developed.'*(Gregory 1981, p43)*

Gregory's claim that technology has moulded the ways that we think is important. So too is his belief that we are only dimly aware of its implications. As we now engage in teaching technology for the first time to all children we shall become more aware of how it has moulded our way of thinking. By this means, our understanding of the immense importance of technology, not only as a means of doing things but as a means of structuring our thoughts, may become less 'dim'.

Archaeology has shown that technology is much older than the human race. It is now possible to consider how the demands which technology made upon the special dexterity of our ancestors, over many thousands of years, would favour the development of the mental capacities which we know as human intelligence.

Teachers of young children are aware that practical activity stimulates the use of language by infants and often generates an eagerness to learn how to realize their ideas. Such teachers should find no paradox in the discovery that hands and tools not only preceded human intelligence, but also contributed to its development. What the human hand and body could do better than those of any other species was handle the material they found in their environment.

Although solitary activity is valuable for a child, mainly because it offers uninterrupted contemplation, there are limits to what can be learnt alone or only in the presence of other children. The presence of an adult should allow for more challenging conversation for the child and opportunities for a teacher to witness learning as it occurs. Children inherit from their ancestors the physical and sensory predisposition to explore and manipulate the materials in their environment. However, since the environment is increasingly structured by technology and since access to materials may vary, the supportive intervention of a teacher is essential.

Children must constantly adjust their knowledge of the world to make it compatible with what their senses tell them about it. Talking with adults, particularly when linked to activity with materials, helps children make that adjustment. This is evident in the ways in which children easily change their view of things to accommodate new experiences or information. Children necessarily use conversations, drawings and models to help them form and modify their ways of thinking. Their knowledge, however it is framed, is lightly held and they are accustomed to adjusting the propositions they make about how things are or how they work.

Because of their limited experience, there is inevitably a gap between what children know and what is known by adults around them. There is a further gap between what adults, in general, know and what exists to be known from the work of experts in all the specialist fields.

This gap in knowledge is professionally mapped and inhabited by teachers. Their responsibilities include meeting the expectations of adults on the one hand and providing the information and guidance required by children on the other. The skill of teachers lies in balancing these responsibilities and mediating between them. The teachers' task has been enlarged since the first national education system of elementary education was set up in the mid-nineteenth century. It was then expected that a teacher, or any reasonably educated person, should know all that a child was expected to know on leaving school. Since then, specialist knowledge has enlarged the pool of information which children have to navigate and within which they must acquire and maintain confidence. This huge expansion of knowledge means that children must become confident of their ability to gain access to specialist knowledge by means of information technology.

The advent of the National Curriculum
Along with the size of the teacher's task, both the possibility and the cost of disappointment have increased. The cost to children can amount to loss of confidence in their ability to learn, with all the loss of self-esteem and apparent loss of respect for authority which may ensue. The cost of

LEARNING THROUGH ACTIVITY WITH MATERIALS

The ease of this child in her environment derives from confidence in her senses. In physique and brain she is much like her ancestors, from whom she has inherited adaptability and curiosity about what comes to hand. However, her mind is already furnished differently from theirs because of the technology and culture within which she is growing up. The teaching she has received has enabled her to gain knowledge and retain her confidence.

disappointment to parents and teachers, who collectively invest in education, is probably incalculable since expectations vary and are a matter of debate. The discussions which preceded the advent of the National Curriculum did much to focus that debate on reasonable expectations of the education system.

The National Curriculum has provided descriptions of programmes of study which together reflect increased expectations but which also identify reasonable staging posts en route. Descriptions of the stages serve to indicate how the journey across the ever-extending gap in knowledge may be undertaken.

In this respect, the National Curriculum documents do much to support the teachers and should help to ensure that the appropriate provision is made to achieve their stated purposes. It is gratifying that specialists, who represent advanced expectations, have described, with the help of teachers who served on the working parties, some of the enabling stages through which children should pass. This particular dialogue between specialists and teachers should continue, for both the 'territory' of education and the way it is perceived will continue to change.

Since design and technology draw upon knowledge which is beyond the ken of any one person, no teacher should expect, or even attempt to know, the whole 'subject'. What can be known however, is the nature of the design and technology work in which children should be involved and the kind of demands it should make upon them. Children and their teachers need to establish the fundamentals of the disciplines from the outset. The attainment targets for National Curriculum Design and Technology set out the framework for these disciplines and in doing so stress the importance of children identifying needs, making proposals, executing their plans, and evaluating and appreciating the outcomes of their own activity and the activities of others.

The attributes described in these attainment targets follow directly upon those which characterized the rise of human intelligence and the adaptability of the species, which were discussed in the early part of this chapter. They also reflect the practice of good designers and technologists today. The ways in which we see and understand the fundamentals of design and technology will change as their frontiers develop: it is important therefore, that teachers in schools, and practitioners in specialist fields, continue the dialogue which has now been established.

Laying the foundations

Teachers of young children have a special responsibility since they lay the foundations of study across the whole curriculum. They also have their own valid discipline and this must be rooted, not so much in attempting to know everything that children may be expected to learn during their school years, but in recognizing and reinforcing the children's capacity to use authentic methods of working. The word 'authentic' here applies to methods which, while being appropriate to children, also genuinely reflect the practice of the best practitioners. To enable children to use such methods, teachers need to balance what they know about children with feedback from the various fields of specialist knowledge. Even so, teachers will always be vulnerable to the charge of not knowing enough, but they must be confident of their own expertise.

The quality of work achieved by teachers in schools and in some advanced courses of professional study suggests that the ground is already well prepared to inform a wider public of the basis upon which teaching expertise is founded. There are many people outside the teaching profession, experienced perhaps in the disciplines taught in schools, with whom a mutually informative dialogue should be maintained.

Early learning techniques
We are not entirely detached observers of the physical world since some of the properties and behaviours of materials are learnt within our own bodies. Watching a child learning to sit up, roll over and crawl we can see how it gradually learns to use its arms and legs as counter-weights and levers. For example, a baby puts considerable effort into learning to turn from lying on its back to lying face down, which is a necessary preliminary to crawling. In early attempts to roll over, the elbow of the downside arm is a frustrating obstruction until it is tucked into the side. This arm then provides a narrower fulcrum about which the body may turn. The main force for the rotation may come from the strong arching of head and neck with final assistance coming from a swinging movement of the free arm and leg across the point of support.

Rolling over when clothed presents different problems, and variations in the slope and texture of floor surfaces offer different degrees of purchase to hands, head, feet and elbows and call for the learning of alternative techniques. The child learns from day to day, repeatedly making similar movements and adjusting them to those previously found to be efficient. As a repertoire of movements is acquired, actions can then be selected for particular situations.

Tentative walking follows quickly upon standing and no sooner are toddlers able to adjust the balancing of their bodyweight to the demands of forward motion, than they are able to lift and transport large or bulky objects held before them.

No doubt many of a child's early movements are reflex actions and the drive to achieve independent movement may be instinctive. Parents and

Materials in design and technology

students can observe that learning is involved and that it varies from child to child. Some extend the crawling period over many months while others move on to walking in a few weeks. By the time they are able to walk, it can be assumed that all children have learnt a great deal about the behaviour of their body mass and their immediate physical environment.

Of objects encountered in the environment, those which provide stable support appear to be quickly distinguished from those which do not. Objects like doors, wheeled toys and pushchairs receive much attention but once their idiosyncratic movement is sufficiently understood they are used with discrimination as an aid to standing. Kinaesthetic learning is not confined to the child's own body nor to materials encountered, but includes learning from manufactured goods and toys from which principles of movement and structure can be absorbed. The occurrence of this kind of learning can pass unseen by adults but the achievements to which they lead, like proficient speech, feeding and dressing themselves, swimming and cycling may ultimately draw high praise.

By the time they are able to walk, children have for some time accompanied their physical activities with oral 'commentary' in a generally encouraging environment in which adult or sibling speech occurs. Basic experience of matter, and some insight into mechanics, seem to be acquired early and without the aid of fully-fledged language. Although they are without words to help 'fix' the concepts, children appear to be able to store and use their experience of balance, rotation, leverage and friction to achieve locomotion in a variety of situation which require them to remember and adapt both their movements and the circumstances in which they are useful.

Adult kinaesthetic learning

This kind of learning, and the kind of knowledge to which it leads, continues to be useful in many aspects of adult life. Kinaesthetic learning is of obvious importance to athletes, dancers, designers, technologists, musicians and artists, and most people can give accounts of learning to use tools, from a golf club to a motor car, and from a spade to a sewing machine. In these situations, although verbal communication and even detailed instruction is necessary to set the scene for learning, the acquisition of mind and body skills does not appear to depend upon

A child moving on a seesaw disturbs its equilibrium. Particularly in the company of others, she will soon learn to balance it and go on to learn how to balance unequal loads. It is in direct bodily contact with materials and apparatus that children first understand how things are and how they work. Physical balance features in apparatus for mathematics teaching and for science, but the concept of balance is essential to many ideas. The technology of the scales gives us a metaphoric vocabulary and a model for justice.

verbal facility. It seems rather to depend upon tuning and refining the awareness of body and limb movements in relation to the materials and circumstances to which adaptation is being made.

Less immediately obvious is the extent to which bodily learning, and the informed perceptions and judgements which accompany it, drive one to use and refine language. Some of those who practise a discipline in which kinaesthetics feature prominently, make their perceptions accessible through language of one kind or another. Examples, in addition to written and spoken language, include all kinds of drawing, music, choreographic notation, mathematics and a range of computer languages. Einstein acknowledged the kinaesthetic origins of his thinking and spoke of his first thoughts as always being muscular.

ANCIENT CIVILIZATION

Having looked briefly at some of our common ancestry and at shared characteristics of young children, both of which influence our response to materials, it is now possible to see how materials influence the wider aspects of culture.

Evidence of the earliest civilization exists in those parts of the world which offered the best resources for sustaining community life: primarily a combination of near-tropical warmth, which encouraged crop growth, and periodic flooding, ensuring the renewal of fertility in the land. These conditions prevailed in relatively small areas of the earth's surface and populations were, by modern standards, very small. Such areas included the valleys of the Yellow River, the Ganges, the Tigris, the Euphrates and the Nile. The distances dividing these areas ensured long periods of separate development.

The most prolonged and evident separation was that between eastern and western cultures, which still has consequences for contemporary society. Amongst the more obvious of these materially-inspired differences are the typical perceptions of reality. Where the West developed the pen, the pencil, standard inks, a restricted alphabet and linear perspective, the East developed the brush, variable inks, fluid calligraphy and atmospheric perspective. Western realism is characterized by delineation of stable forms and entities, whereas Eastern realism is characterized by space, movement and changing forms.

First, the very existence of these cultures and second, their ways of thinking, depend on material circumstances and the technologies that they could support. We associate Egypt with two dominant material features: the waters of the Nile and massive stone. From these came a highly developed community and a great hierarchy sustained by monumental, tangible symbols of royalty and deity.

In contrast, Mesopotamia had no stone, but from its abundant alluvial clay, developed bricks and clay writing tablets. While brick-built walls and ziggurats lacked the permanence of pyramids, their softer properties gave rise to the lasting fame of Babylon's Hanging Gardens.

Material dependence

Every early culture appears to have made the attempt, through myth and legend, to explain its dependence upon material resources and to propitiate the powerful influences which they exerted. With agriculture and herdkeeping as the prevailing technologies it is not surprising that these materials and activities were used to describe the personalities of the gods. The practice of conveying religious teachings, by using models drawn from agrarian concerns and skills, has survived the Greek, Judaic and Christian traditions, to the present day.

In more contemporary times, particular materials came to be in demand because entire technologies depended upon them. The great fleets of the seafaring and trading nations were wholly dependent on a constant and convenient supply of wood with which to build their ships. The availability, for example, of timbers from adjacent forests contributed to the eminence of Venice at sea. When these forests became exhausted, from the building of a battle fleet, neither the environment nor the fortunes of Venice were to recover.

Further ravages upon the forests of the Mediterranean were wrought by goats which moved westwards from Turkey. David Attenborough writes:

> 'The Turks, who had never practised irrigation with the skill and dedication of the Arabs, had brought great numbers of goats with them into their newly acquired territories in the eastern Mediterranean, and now the herds spread all over the Levant and Greece. Once goats are established, the land stands little chance of recovering its trees and regenerating its topsoil. The goats consume every seedling that sprouts and every leaf that unfurls .... So the land remains barren. The woodlands that once ringed the Mediterranean and provided a home for a rich population of animals had now largely gone.' *(1987, p173)*

Historically, it can be upon such unpredictable or uncontrollable changes of circumstance or ecology that cultures may seem to be arbitrarily raised or cast down.

The technology of shipbuilding moved north where essential timber resources remained plentiful and where the centres of shipbuilding thrived for the next three hundred years. The northern shipyards were again to be fortunate when the industry transferred to building ships from iron which was also readily available to them.

The history of the oak
If Venetian fortunes had suffered from the goat, Elizabethan shipping benefited from the pig. The ships of English oak owed their source of timber to the fact that the Saxons had grown extensive oak forests to supply their pigs with acorns. Oak is a hard, homogeneous and strong wood, stable when wet and therefore particularly suitable for making keels, decks and ribs for ships. Remains of the oak go back nine thousand years in England, where it survived as a forest tree until cultivated.

> 'The pedunculate oak was the tree of the pig-pasturing Saxon, great builders in wood, and it persisted at the heart of our tree-based economy, until ships of oak became our "wooden walls" *(Evelyn 1664)* and "hearts of oak were our men" *(Garrick 1770s)*. The trees were open-grown to get the spreading crooked shapes for "cruck" houses and ships' "knees". Then very quickly in the 1830s and 40s, ships began to be made of iron and the picturesque trees were left to stand in the Royal forests, bearing a heavy aura of ancientness. Of course, oaks were venerated, preserved on village greens and at crossways, preached under, hidden in by fugitive kings.' *(Wilkinson 1987, pp34-35)*

In this brief paragraph Wilkinson first outlines the usefulness of oak as a naturally occurring food source and then as a tree cultivated for the properties of its wood. He goes on to show how, as its use in shipbuilding declined, it rose above utility to become a metaphor for patriotism and noble sentiments.

The case of the oak tree may serve to draw attention to the links between materials, technology, language and ideas. An available material is put to use and becomes familiar; new uses develop and may give rise to complex technology; this in turn leads to new structures, to new ways of thinking and new language. Because of its importance to teaching, it will be necessary to return repeatedly to this theme.

2 NEW AWARENESS

FEWER people now work directly with materials, which may mean that opportunities for individuals to acquire knowledge of them vary more now than in previous generations. It is important, therefore, to consider some factors which may influence our perception of materials and our awareness of their importance.

CONSEQUENCES OF SPECIALIST EMPLOYMENT

It is characteristic of our time, and of western culture generally, that we are nearly all employed as specialists. We are not self-sufficient; we depend on others to supply most of what we need to maintain our lifestyle. Most of us contribute what our education and specialist training have equipped us to offer and we earn the means to purchase the goods and services offered to the community by other specialists. Consequently, many of us are distanced from direct evidence of our dependence upon material resources. However, there is a small group of specialists who do work directly with materials and have reason to be more acutely aware of the importance of material resources. In this way, specialization contributes to the wide variation in the importance which individuals attach to materials.

In some primitive societies, the influence of material resources is universally apparent because the users live close to their sources of supply. In such a culture the need to seek grazing for the animals, for example, may lead to an essentially nomadic way of life, but at times it may be necessary to stay near sources of earth materials such as metal ores or clay, and within reach of the fuels which are required to utilize them. Activities might then be dictated by seasonal changes.

In complex societies – those which are thought of as more advanced – what is made by one person or group is exchanged for what is made or provided by another. Sources of supply, and the nature of the material govern the location of industry and the technology which can be developed there. Consequently, there are a number of individuals, families or small groups whose location, lifestyle and status become identified with the technology they provide and the resources they use. The presence of many such communities contributed to Britain's economic growth during the nineteenth century.

Regional specialization
Following the Industrial Revolution and well into the present century, the UK developed many communities and cultural traditions founded upon regional crafts and industries. In England these included: the Staffordshire potteries; Sheffield iron and steel making; wool and cotton spinning and weaving, west and east of the Pennines respectively; North Sea fishing; Yorkshire and Midland mining; and shipbuilding in the North East. Similar specialist communities could be found in Scotland, Northern Ireland and Wales. Local sources of materials and fuels, and particular geographic conditions favoured the development of these local industries during what was an extended period of technological innovation and industrial expansion after the Industrial Revolution.

The Carrara quarries have been worked for 2,000 years, and provide an example of a huge but finite world resource. Some 300 quarries dominate the lifestyle and employment in this area of Italy. The most celebrated piece of Carrara stone was that from which Michelangelo carved his statue of David, commissioned as a symbol of the independence of the Republic of Florence.

The location of materials, together with their associated working methods, standards and skills, enriched the character of communities and contributed to their identity within the nation. What began as resource-based specialization in the regions became sources of reliable skills and standards. Regional specialization also allowed other areas to develop as agricultural, commercial, military or academic centres, with their own perceptions and values. Some of these traditions were very strong and continue to exercise some influence on present-day attitudes. Examples are to be found in the many variants of harvest festivals, works carnivals, regimental parades and inter-college sports, which provide ritualized expression of deeply-felt loyalties.

Awareness of material resources, either as a basis for a lifestyle or as a factor in economic activity, varies from region to region. Teachers who move from one area of the country to another often speak of these variations as expressed in attitudes to work and in the aspirations and expectations which children appear to inherit. Knowing that different attitudes and perceptions may have deeper roots than mere prejudice can help us become more sensitive to such values and therefore more capable of usefully extending them.

Necessary adaptations
With the growth of markets and the internationalization of industry in the latter half of this century, the traditional, lifelong dedication of communities to one major industry has undergone a dramatic change.

Where whole communities were once predisposed to hand on the values and skills of their parents' culture, altered circumstances now place a higher premium upon diverse, more adaptable skills and a wider understanding of technology. Adaptation is most difficult in highly specialized communities where some of the finest craft traditions exist. Where schools face the task of encouraging new awareness and adaptability in children, it is clear that this may best be achieved by maintaining and extending their respect and understanding of their community's traditions. Teachers may find that a long-term view of technology is helpful in enabling children to see the relevance and importance of traditional values as they learn to adapt to the accelerating pace of change.

Transport and food
The technologies of transport and food have liberated most of us from the need to live close to our work and we are thus able to maintain a particular lifestyle without living close to certain sources of materials. Transport, itself a heavy consumer of fuel resources, has contributed much to this freedom, but it may have made society less conscious of its dependence on the continuous availability of particular materials. The many processes involved in food technology, from curing and cooking to canning and refrigeration, has liberated us from the dominance of the seasons and even from the worst effects of drought and crop failures. The combination of improved transport and food technologies means that even local supermarkets are able to supply a wide range of fresh food, with little variation in quality throughout the year.

On the whole, the consumer has no direct involvement in the processes which maintain a staple diet and a customary mobile lifestyle. We may know where our family car was purchased and possibly in which country it was assembled, but we are not likely to discover whether the materials were derived from scrap by local reclamation or smelted from imported mineral ore. It is part of our liberation that we leave such questions, and those regarding farming and food production, to the experts who understand the processes which maintain our supplies.

TECHNOLOGY: A CURE FOR ITS OWN ILLS?

It may seem that technology, although founded upon the properties of materials, renders us more ignorant of our continued dependence upon them. During the last hundred years technology has enabled us to consume more of the world's resources without personal involvement in the consequences. Not only is our own consumption increasing but more of the world's population wish to enjoy the benefits of the consumer society – ironically by selling their own material resources. The complex lifestyle which technology engenders tends to obscure the extent to which our culture is still shaped by the nature of the materials upon which both it and its technologies are founded.

However, technology must be credited with new public awareness of the consequences of over-consumption of energy, food and material resources. Space technology has made it possible to see the earth as a whole, while television has provided intimate views of the needs of the peoples, animals and plants who share the planet.

Television is a technology which exploits the electrical behaviour of materials as they change under varied intensities of light. So our visual awareness has been immeasurably extended and it is now possible to observe, not only hitherto unknown creatures in their habitat, but also how distant habitats are affected by the products of our culture. Technology, which for a time seemed to cushion us from the effects of our action (or inaction), now provides the means of making us face up to them. By constantly offering new structures for our perceptions, and new possibilities for action, technology demands that we do not act without first considering possible consequences.

Responsibility for choice
The word 'design', in the context of this discussion, may be applied to any course of action which is selected from possible alternatives and undertaken as a result of anticipating their outcomes. In the past, physical versatility and its extension by the use of tools, imposed upon humans the need to exercise choice, and discrimination between alternative possibilities required perception and imagination. These abilities, first to perceive an existing state of affairs and second to imagine how an action might change it, are prerequisites for any design process.

It is important for teachers to understand the significance of perception and imagination in the development of design capability. Critical and planning abilities are also important aspects of design, but attempts to develop these without sharpening children's perception and alerting their imagination are counter-productive.

Any advance in technology widens the possibilities for action and therefore for choice. Technology cannot advance without some choices, conscious or otherwise, being made about possible outcomes. All the choices, which enable us to use technology, contribute to what is meant by design. It follows that the quality of attention given to design governs the quality of the technological outcomes. The earliest manifestations of design enabled individuals and groups to survive by learning to make appropriate decisions. Recent advances in technology have made us aware that future action must be designed not only for the survival of the species but also of the planet we inhabit.

New developments
The responsibility for good design which accompanies the benefits of technology is exemplified by recent developments in biotechnology. These enable us to modify the genetic structures of living material. This huge advance upon selective breeding requires a new awareness of our responsibility for the well-being of life on earth and calls for appropriate utilization of living and non-living material resources. It is more evident to the present generation that the welfare of the human race, and of most other species, depends on humankind's efficient management, care and renewal of world resources.

The personal qualities required to meet the demands of such a responsibility cannot be easily determined or described, but the attainment targets for capability in design and technology within the National Curriculum are clearly relevant and likely to be of increasing importance as the nature of the demands becomes clearer. The effect of our new awareness of this responsibility, at present, does not so much influence the content of education in design and technology, as underline its purpose.

Another recent development in technology, of which teachers need to be aware, is an increasing capacity to 'invent' materials with new characteristics. The first such invention probably goes back to the Bronze Age, or earlier, with the addition of hair to dung or clay. What is new is the rate at which new materials are coming forward and the sophistication with which their properties can be planned.

New materials enable us to undertake different tasks and create new structures, and most households contain examples of new synthetic fibres and plastics. While they obey the same physical laws as other materials, synthetics can be made to behave like textiles or metals and can be presented in opaque or translucent forms. They can be endowed with properties like flexibility, rigidity, resistance to heat or corrosion, or strength under compression or tension, and so can now be designed for a particular purpose. Offshoots from this technology include a new interest and understanding of natural materials and additional respect for the qualities inherent in the materials and structures of living organisms.

One such example is the development of human-powered flight, a concept which is at least as ancient as the Greek legend of Icarus (which places him some four thousand years ago) but which was not fully realized until 1977 with the synthesis of carbon-fibre structures and strong polyester membranes. These have renewed the admiration for nature's ingenuity in the evolution of materials and mechanisms which permit flight in a variety of forms, from seeds and insects to birds and mammals.

Architecture, engineering, clothing, communications, furnishing, shopping, travel, sports and recreations have all felt the impact of synthetic materials. Fulfilment of the potential of these materials depends on good design, both of the materials themselves and of the end product.

Finally, we should acknowledge a growing understanding that language, and therefore education in language, is rooted largely in sensory discrimination. Materials activate the mind and every new technology provides new ways of seeing and offers new models for the construction of ideas. An important reason for the inclusion of purposeful, practical activity with materials in the primary curriculum is the demand it makes upon language and the structural and syntactical experience it provides.

ABSORBING NEW INFORMATION

Gaining competence
The act of selecting and assembling things in order to convey meaning, or to present ideas, has much in common with the use of language. Both provide a child with opportunities to adjust new knowledge to fit into their scheme of things. While education is undoubtedly concerned with imparting information and increasing children's awareness of what exists, it also includes enabling them to act upon their knowledge. Awareness, without the competence to act upon it, must logically lead to fear of taking action and to anxiety about further learning. Teachers need to be aware that both adults and children can deal with such anxieties by preferring not to learn.

The human capacity for absorbing information is enormous and the natural way for this absorption to occur is in active contexts which give the learner opportunities to relate new information to their existing store of knowledge. Activity itself allows existing perceptions and thought to be adjusted and competence to be maintained, particularly where teaching supports activity. Anxiety may arise when a learner is unable to relate new information to existing knowledge which has hitherto provided an adequate basis for action. Opportunities to modify old perceptions and to test new ones can dispel such anxiety, but if anxiety is not dispersed the learner may find no alternative but to reject new information so as to remain within the security of previous competence.

Technology advances at such a rate that there is always scope for anxiety. Different interest groups may have disciplines which are founded on the use of the same materials, but have developed traditions along different lines. Because of this, it is not unusual for specialists to have conflicting views: architects with builders, dressmakers with costume designers, lighting experts with scenery painters, and so on. Cooperation between specialists is best achieved where they are able to agree over the particular requirements of the work in hand.

Ezio Manzini *(1989, p44)* points to the inherent difficulty of making information about new materials available because the language of esoteric research has not previously lent itself to the demands of coherent communication with the wider public.

Responsibility for communication
Similar problems exist in education. Schools are themselves specialized institutions and teach by means of a subject-based curriculum. Communication between individuals inside schools and those outside needs to be matched by effective communication between the different interests represented in the curriculum. Because design and technology both contribute to and draw upon the whole curriculum, they should provide a useful focus for cooperation.

In the wider community the sheer scale of responsibility has increased. It is not merely domestic or local as it was in pre-industrial economies nor is it merely within an extended family, or regional industry. While it continues to embrace all of these, even the single matter of responsibility for resources is vastly widened by technology and requires an extension of awareness and cooperation. Responsibility is now international and global. Technology will, no doubt, contribute to such an extension by freeing some human resources from repetitive labour and by facilitating the transfer of information. How well such developments are used may ultimately depend upon our capability in design and the extent to which we develop this capability in our pupils.

3 TRADITIONAL PRACTICES

HUMAN RESPONSES TO MATERIALS

ARTISTS, craftworkers, scientists and technologists work with materials for different purposes and according to different criteria, and all are disciplined by the traditions within which they work. Our purpose here is to demonstrate that these traditional disciplines are rooted in human responses to materials.

It is important to acknowledge that no person, whether artist, scientist, craftworker or technologist is exclusively that. Each specialist occasionally assumes aspects of the other traditional roles quite naturally. Children act in the same way: sometimes displaying the objectivity of the scientist or the pragmatism of the technologist, while at other times exhibiting the sustained care of the craftworker or the acute perceptions of the artist. These characteristics in children are indications of potential rather than evidence of readiness to work within an established discipline. Being able to recognize them as they occur in children's work enables teachers to offer authentic challenges and appropriate encouragement.

The most ancient response to materials technology seeks to utilize the particular properties of available materials. For example the flexibility of a sapling can be exploited to apply tension to a length of hair or fibre to make a bow. The 'cord' of such a bow can, while under tension, be used to turn a spindle or to generate a sound when plucked, or when released can be used to hurl an arrow or to spring a trap.

Different responses
Given a technology, a characteristically art-like response is to explore its potential for the development of new forms or of new perceptions of

Materials in design and technology

These paintings by infants show different perceptions of cherry trees blossoming in the school garden. The form of the tree observed is complex and it is evident that the children have made different propositions about how its form can be characterized. These propositions have been influenced by the materials used.

existing ones. In the case of cord under tension, new possibilities were afforded by the straightness of the cord and the regularity of the arc of the bow. Both were important visual models for many subsequent ideas, such as the long-standing representation of the flat earth surmounted by an arc of the sky. Another art response might be provoked by the sounds made by plucking the string of the bow. Technology has gone on to provide many alternative ways of making new structures and new sounds which have been exploited in sculpture, architecture and other art forms.

A craft response, the urge to make well what is possible or required, has figured prominently in response to technology. In the case of the bow, whether as an instrument for music or as a weapon for hunting, the quality of its performance would reflect the quality of the making. Great craft traditions were to develop both the functional and symbolic quality of the different kinds of instrument making.

The science response has sought to understand the behaviour of materials and the rules which underlie their structures. The science response to the bow would seek to understand the variables affecting the efficiency of the bow, and the relationships of the sounds. Science has made great advances and provides information and understanding which has allowed art, technology and craft to proceed with greater awareness of the factors which they each seek to control.

The traditional disciplines of technology, art, craft and science are founded upon a range of responses to materials. These are not arbitrary; first, they are characteristically human and second, they are structured or influenced by the nature of the material involved. Education for design capability must provide what amounts to an apprenticeship in each of these disciplines, building, during the primary years, upon the natural responses of the children. If each of those disciplines is examined, the following points can be observed:

- The technologist is concerned with the properties of materials primarily to understand how they may best be related to processes which facilitate what we are trying to achieve. The apprenticeship of a technologist must provide many opportunities for refining procedures toward desired outcomes.
- The craftworker is concerned with the properties of materials in order to understand how their qualities can be used to form and enhance the quality of the product. What is made well must, in addition to functioning as intended, give satisfaction to the user through all the constituent qualities of its form, colour and texture. The apprenticeship of a craftworker must provide many opportunities for making judgements and acquiring skill.

- The artist is concerned with the properties of materials in order to understand how their qualities may be used to structure perceptions. The apprenticeship of the artist must provide many opportunities for relating the properties of materials to aspects of both real and imagined perceptions.
- The scientist is concerned with the properties of materials in order to understand how these are a function of their composition and structure. The apprenticeship of the scientist must provide many opportunities for observation, prediction and experiment.

Capability in design and technology needs to draw on each of these traditions and each kind of 'apprenticeship' should be a part of children's sustained experience.

DRAWING

There is no single way in which children learn to design: on the contrary, there are many ways but it is significant that the act of drawing plays an important part in all of them. The word 'design' is itself derived from the word for drawing and particularly for making a mark. One reason for the use of drawing in design is that drawing materials come readily to hand and are amongst the first over which most of us gain direct control. All that is required is a surface and a material to make a mark upon it.

Pencils, as we know them, made from a mixture of graphite and clay, have been in use since the end of the eighteenth century. Since then we have seen the arrival of the continuously-flowing fountain pen, the ball-point pen and the felt-tipped pen, to name but a few of the markers which influence our drawing styles. Before the advent of the pencil, drawing materials included charcoal, chalks, inks from many sources including the sacs of squids and cuttle fish, and a variety of metal points, quill-pens and brushes. Surfaces used included wood, skins, bone, stone, textiles, metals, and a variety of hard and soft, smooth and textured papers. Each of these instruments and surfaces has in their time exerted an influence on how the world is perceived and thought of.

Drawing is a way of thinking about and expressing what we see, imagine or feel. Drawings themselves are not only the product of human movement, but also a trace or memory of that movement. When we draw, as we move our hand we see a mark corresponding to the movement we have made, just as when we speak we hear what we say. With practice we learn to understand what we think by listening and making corrections to what we say. In the same way, designers learn to see in their drawings evidence of how they think, and drawings often gain vitality and meaning by having second thoughts built into them. Designers draw not only to demonstrate their thoughts but also, by a process of trial and error, to gradually reduce the possibility of mistakes.

Representational marks

In order to make a drawing we need to propose that a mark can stand for something else. Every child who draws makes that proposition. What seems to be important at first is recognition by the child that one thing, a mark, can be made to stand for another, for example a person. The child may then give to a mark the familiar name of a parent or friend. Modifying marks to represent particular aspects of a person comes later.

The child who draws confidently and the adult who draws successfully both learn from experience that the marks they make at any one time do not have to resemble the object they represent. The meaning of each individual mark is developed only as other marks are made in relation to it: in this way a set of relationships is made which is compatible with those observed in real-life situations.

There is, however, another important kind of meaning in marks which comes from the way they are made. The effect of whether a mark is made quickly or slowly, with great precision or vigorous abandon, is visible in the mark. In this way, children's drawings acquire meaning from the qualities of feeling and gesture which produce them.

LEARNING TO SEE

In order to understand how materials influence perception we need to understand that 'seeing' has to be learnt. Herbert Read describes the whole history of art as a history of ways we have learnt to see the world:

> 'The naive person might object that there is only one way of seeing the world, the way it is presented to his own immediate vision. But this is not true, we see what we learn to see, and vision becomes a habit, a convention, a partial selection of all there is to see, and a distorted summary of the rest.' *(1959, p12)*

Read draws attention to the fact that, having learnt to see in the way that we do, it is difficult for mature people to imagine that there is any other way to see. Like learning to walk or swim, seeing becomes so natural that we are hardly conscious that learning was involved.

The inverted images of the world, which lie upon the retina inside each eye, are constantly changing. The images *per se* carry no explanation of the world to which they introduce us: we must supply this for ourselves as we discover continuity and recognize consistencies within the images. As we rise from a chair and walk across a room the retinal image of everything in our field of vision undergoes a series of continuous changes. It is from such myriad images that we have to learn to see, that is, to construct for ourselves a constant world.

Colin Blakemore explains that it is necessary to learn to distinguish between retinal images and the constancies to which they refer:

'As you watch a man walk away, you see that he remains the same size (although the image of him on your retina certainly shrinks). Look at a coin when you turn it in your hands and you perceive it always as a round disc of metal (although its image is an infinite series of elipses). What we reconstruct within our mind's eye are the constant physical properties of objects; and so we should, because those are the things we need to know.' *(1977, p67)*

Using other senses
However, the data which we use to build up our perceptual models is not confined to that obtained by means of the eye but may draw upon all the senses. What is seen is confirmed, or qualified, by touch and taste. The potency of sense images acquired in childhood is illustrated by the strong memories of particular childhood events which can be triggered for adults, many years later, by evocative smells or sounds.

Just as we learn to make comprehensible images from light impulses on the retina, so we learn to construct and expand our perceptions by means of impulses from all the senses. Just as learning to 'see' a moving ball involves interpreting the changing retinal image, as an object of constant size, so learning to catch a ball makes the further demand of relating the new sensation of its impact upon the hand to the familiar feeling of holding it. Our sense of touch, which involves not only the hands but the whole body, brings an extensive range of sensations which we learn to relate to each other and to those from other senses.

It is from such data that we construct a world of identifiable objects with constant physical properties. It is this world which we must successfully inhabit and share with others.

Perceptions with and without words
Learning to furnish the world with named 'objects', however, is only one of the most basic of perceptual efforts. All kinds of 'entities' have to be repeatedly encountered until they are recognized. Conversations between teacher and child strengthen the ability and confidence of the child to perceive, name and use objects. Adults have constructed many ways of perceiving time so that we can, for example, refer quite truthfully to the same event as having occurred 'last year', 'six months ago', or 'the term before last'. Each description uses a different means of parcelling time into comprehensible units. Adults have learnt to handle models of otherwise intangible entities like time, but children must try them out.

Sometimes children become familiar with a word before they have identified the entity to which it refers. At other times perceived entities may appear long before children discover the words to describe them.

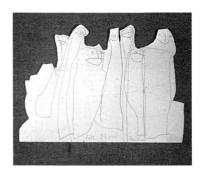

The top picture shows how a four-year-old child depicted individuals in her family and then used her new skill with scissors to cut them out and play with them in real space. The bottom picture shows how the scissors have been used to follow the perimeter of a group of bottles on a wine merchant's greetings card. The child then proceeded to draw, on the reverse of the cut-out greetings card, a group of people she called 'daddies'. This example shows how the form of a material can help to generate an idea.

Learning to see with others

No one is alone with their senses. Our culture provides us with many clues to help us establish orderly perceptions. For children these clues include all kinds of pictures and models; amongst these are the important images which they themselves make and the models they construct from available material. Education must challenge children to test these clues against their own perceptions and those of their peers. In the long term it is desirable that the perceptions of individuals both enrich and become compatible with those of the culture to which they belong.

The process of modifying our perceptions in the light of our own experiences and those of others is our education and is not confined to childhood. Language is the greatest aid in this process, because to extend one's vocabulary is, necessarily, to extend one's perception.

MATERIALS IN PERCEPTION

Light and sound, feel and taste, are all products of matter, and their characteristic properties govern the way we perceive them, as well as how we can use them. Light travels fast, and does not deteriorate over long distances; this property allows us to react quickly to distant events. Sound also travels, but less quickly, and it deteriorates markedly over distance; nevertheless, for those with acute hearing, its wide variability allows it to carry a vast range of signals and warnings. These characteristics have rendered sight and sound useful media for the construction of language.

Smell travels slowly but diffuses rapidly. It helps us to confirm recognition of important things when they are close to us, such as whether food is good or bad or, for a child, the identity of a favourite possession or the reassuring presence of a parent. Because of its intimate nature, smell can be an essential clue to imminent danger and is therefore important during the vulnerable period of childhood. Taste and touch are directly physical and require material contact of some kind, though the body can also detect vibrations below the level of sound, and other clues which are used to assist the perception of motion and position.

The special receptor organs which enable us to perceive these clues are also limited by their own material characteristics. The human eye, for example, has evolved to the point where it functions well for the species. The materials of the iris and lens admit a selected range of the total light spectrum and focus it upon the retina, where specialized cells respond to stimuli and provide the brain with the signals it requires to form its images.

The other organs of perception – the ear, nose, fingertips, and nerves distributed throughout the whole skin surface – are physically structured to detect different stimuli and to convert them into useful signals to the brain.

The use of a number of senses together to modify and refine images depends upon our learning to recognize particular qualities and forms encountered in materials and in materially composed objects. This ability to recognize and relate different kinds of sensation is a necessary condition for language and is, in turn, enhanced by language.

In learning to see, in terms compatible with our parent culture, we are assisted not only by language but by all the material structures, representations, pictures and models with which that culture surrounds us from birth. The traditional nursery accessories, such as mouthable toys, rattles, musical boxes and picture books are all designed to provide the means of activating the mind through the senses. Their constancy, among so many changing factors, is important to the child as a basis for learning to recognize the properties of material objects, and for learning to see.

LEARNING TO READ

As we learn to see so we act. The construction of visual and tactile images in the mind is clearly facilitated by language. However, in order for language to assist us 'to make sense' of our experience, we must have a sufficient repertoire of sensory images for language to call upon. When we look at a passage of writing, our eyes record the printed page but what we see with our mind's eye is that which we reconstruct from our experience of phenomena like those which the words describe. We call the means by which we proceed from the identification of symbols to the construction of images in the mind, 'reading'.

The same principle applies to looking at drawings. We can choose to see drawings as lines and smudges of graphite upon the paper or we can read them as the events or objects which the marks depict. A further stage of this process applies to reading, not just drawings or works of art, but any objects which have been made. We can choose to see, for example, a chair as something to sit on, or as an object which incorporates in its material form the perceptions, imagination and skills of its makers. It is this further extension to reading which is necessary for the appreciation of design and for the development of design capability.

The repertoire of images which is necessary for the function of language, consists of those which we have learnt to construct from our sensory acquaintance with the material world. Herbert Read goes on to say:

> 'We see what we want to see, and what we want to see is determined, not by the inevitable laws of optics, or even (as may be the case in wild animals) by an instinct for survival, but by the desire to discover or construct a credible world.' *(Read 1959, p12)*

CONSTRUCTED IMAGES

For Read, the materially based art of any period was the means used to construct reality. Certainly, throughout history, the ways in which humans depict the world has undergone some profound changes, strongly influenced by the media used. We are accustomed, in the present time for example, to seeing in our homes instant moving pictures of world events, like wars, whenever they occur.

In order to understand the significance of television as a medium it is necessary to imagine how perceptions of the world were influenced in the past, when the only media were much more selective and limited. What kinds of mental pictures coloured the behaviour of someone living in fifth-century Ravenna, for example? There were no images of figures with articulated limbs or differentiated proportions; no images of atmosphere, weather or varied conditions of light; no instant news pictures to confirm or deny rumours. For nine hundred years images depicting heavenly hierarchies and earthly orders were the information technology of their day.

Learning to see has, for each generation, meant learning to see in terms of the culture of their own time, since this provides the clues to a stable and shared reality. A considerable effort of imagination is required to appreciate the shock of the new realities, depicted in fourteenth-century Florence by Cimabue and Giotto when not only did figures begin to have expressive limbs and faces, but clothing began to hang as though subject to the force of gravity, and the earth itself appeared to be subject to change. The importance of making this effort of imagination is to recognize that perceptions change from one generation to another and from one material medium to the next.

Limitations of news media
In the middle of the nineteenth century, with photography in its infancy, there were no cinematic images of the Crimean War. However, the written despatches of W H Russell, the first war correspondent for *The Times* were no less convincing and influential: his accounts of conditions inspired Florence Nightingale to undertake her famous mission *(Encyclopedia Britannica 15th Ed, Vol 26, p477)*. Newpapers carried drawings from Sebastapol, Inkerman and Balaclava which were reproduced by means of highly detailed hand engraving on wood, a practice which continued throughout the times of the Indian Mutiny and the South African Wars. Characteristically, these wood engravings presented heroic images of brave deeds, with the artistic engraver revelling in the details of the soldiers' equipment, uniforms and horses to which their medium and their skills were well suited.

Photography and film made the greatest contribution to depicting the events of the First and Second World Wars, but during World War II the new medium of radio provided instant reporting of even greater significance and the BBC became the main source of information and entertainment throughout the UK.

In the more recent past, television reporting has opened up the conduct of war to a different and wider public perception. Such has been the growth of the international television public, that strategists now openly discuss the possibility that battles for public opinion may matter more than battles for territory. From this recent evidence it is plain that changes in the media of communication have influenced our perceptions of wars and generally altered the basic structure of the news.

We are aware that edited visual images, accompanied by comment, can appear in our homes soon after any public event. The effect of such images can be so powerful as to be inseparable from subsequent accounts of events. Therefore, it is an essential feature of media education that it should lead to an understanding that every medium imposes limitations upon perception and that reality goes beyond any account which can be given of it.

The benefit of limitations

Just as the characteristics of materials provide structure for craft and technology, so they provide structure for our perceptions and for our ways of depicting and describing reality. Language, for example, is limited to a sequential syntax. For this reason it both requires and enables us to perceive the world by attending to it in sequential units. The speaker or writer directs attention not only to the phenomena described but assigns to them a sequential priority. It is primarily this characteristic which gives language its immense power.

Pictorial media, however, function in a different manner. Although a picture may be painted sequentially, we are not required to read it that way. Furthermore, the limitation, for example, of charcoal as a drawing medium, both requires and enables us to ignore our experience of the world as consisting of coloured objects. This may be turned to advantage in drawings where movement and atmosphere are shown in arrangements of light and dark which overshadow any outlines or separate objects. Clay, as a modelling material, may require concentration upon form alone: it enables us to ignore colour, named objects, or the patterns of tone. The essential point to be made here is that the very limitations of the material we use as our medium provide the means to describe reality. This fact is the key to understanding the educational value of work with materials.

Critical appreciation of all media must include awareness of the

When children draw they use the medium to come to visual terms with what they see and imagine. The form of the drawing varies according to its purpose. Here we see a sectional drawing, like that of an engineer, to show the interior of the Trojan Horse and the film-like device showing the sequence of events.

essential 'otherness' of reality, since to confuse reality with any representation of it is to obstruct imagination, stultify appreciation and suspend criticism. We see this in the stereotyped images which inhibit the development of drawing in some children and in many adults.

LEARNING TO BE CRITICAL

The household medium
In contemporary society we depict our world largely through photographic video-recorded images which show movement, colour and sound. This is executed so convincingly that the images are perceived, not merely as images, but the reality that they describe. It is important to remember, however, that earlier generations were equally impressed by films which now seem unreal: black and white images, still photography and moving pictures strongly influenced perceptions, personal appearance, interior design and lifestyles for many years. The limitations, of which we are now conscious in old films, were hardly evident to the intended audiences, so we in turn need to become aware of the unreality of the images presented to us in modern television. They are, for example, odourless, do not show temperature, are not stereoscopic and do not properly depict space or the effects of fading light. They are limited in the resolution of detail and are restricted both in point of view and field of vision.

More particularly, because the images are so convincing, we lose 'sight' of the fact that the picture on the screen required the presence of a camera operator and crew, with all the intrusion upon reality which that implies, as well as the inevitable series of editorial intrusions.

Television has been in evidence as a popular household medium for less than 50 years and colour television for less than 30 years. The cameras, transmitters and receivers are products of a very competitive international industry founded upon the science-based technologies of optics, materials and electronics. The programmes themselves, whether they are news, education or entertainment, absorb the further skills of designers, artists and writers in addition to journalists, actors, personalities, presenters and teachers. The products of such sophisticated handling are in no sense natural images of the world and therefore need to be viewed by an aware and critical audience.

Television is a young industry and therefore a young medium for the construction of ideas and images. The fact that it can present its images so convincingly should not dull our sensitivity to the fact that, like all preceding media, its limitations contrive to colour its messages.

It is important to be in a position to take these considerations into account when assessing the veracity and usefulness of such persuasive images. Children, unless they are taught to do so, may never learn to take the necessary step to separate reality from received images of it. Hands-on use of materials as media for constructing ideas, and of media technology for storing, retrieving and communicating them, is essential for the development of critical awareness and confidence.

Materials continue to influence our perceptions and lifestyles even in that most sophisticated of environments, the city.

Identifying signs
If we look at traffic as an aspect of urban life we may see that survival depends not only upon our having learnt to see and to read but also upon our having learnt to trust our own perceptions. In learning to see with others, within terms of our culture, some compromises are made but it is important that this education does not entail loss of confidence in one's ability to see for oneself. Survival in traffic, as in all practical activity, requires a keen perception of the materials involved and the limitations of the tools selected.

Technological devices and symbols constructed from materials compete for our attention in every high street. In their form and colour, in their capacity to reflect or to generate light and sound, to simulate movement or to move, they use every known device to deliver and to assert the importance of their messages. Together they are designed to convey the variety of information we need to survive in traffic, for the conduct of commercial and business affairs and to direct us to food and entertainment. All exert an influence upon us and many are designed, quite specifically, to change our perceptual habits towards a favourable view of a product or a direction in which it is proposed that we should travel.

Surviving modern life
Where London has its green man/red man traffic light, New York has its walk/don't walk sign. Neither system is fool-proof and we do not depend upon them entirely. Survival continues to demand that we use our own perceptions to monitor moving traffic, the condition of road surfaces, ambient sounds and smells. Pedestrians, cyclists and car drivers all need to become highly tuned to pick up different kinds of clues. Smell is important to cyclists, for example giving warnings of road works and of altered and threatening road surfaces, which may present hazards for them but which can be ignored by the car driver. While cyclists can change their minds and weave through traffic, motorists must both see and plan ahead if they are to find and transfer to the traffic stream of their choice.

Traffic is used as an example to demonstrate that even in modern life, survival depends upon each of us accepting the responsibility of tuning our awareness to those signs and perceptions which enable us to function.

BEYOND IMAGES

It is for this reason that we cannot depend entirely upon received images. Each of us senses the world in ways which are unique to our position in it. Our position is not static, however, and the picture of the world that we each construct has to enable us to function alongside others. It is necessary to ensure that our perceptions are at least compatible with those of others, and to do this we need to be able to respond to images made by them.

Much of our learning is concerned with understanding and cooperating with the behaviour and perceptions of others. Images produced within a culture can facilitate the sharing of perceptions both within and across cultural boundaries. In practice, the successful sharing of perceptions is a matter of balance. Because we have so much common experience, many of our perceptions may be quite similar. The differences tend to be important, interesting and illuminating; but to enjoy and learn from them presupposes that we are confident enough in our own perceptions not to feel them threatened by different ones. Maintenance of the balance, in children, between openness to the perceptions of others and confidence in their own, is a longstanding challenge for education and particularly for the teaching of design.

We know something of how others see because of the way they act. We, ourselves, act according to what we see and we learn to see in the company of others, and share our language with them. Through language we seek to share our perceptions. There is a sense in which unique perception tends to isolate us, while the sharing of perceptions reduces isolation and provides a basis for community action. Neither the world, nor our view of it, can remain static and it remains a function of materials and media in the hands of artists, designers and technologists to demonstrate the value of new perceptions, new forms and new technical possibilities.

Symbolic behaviour
Susan Langer *(1942)* believes that the real key-note of language is 'the tendency to see reality symbolically' and that symbolic behaviour, rather than anything as sophisticated as rudimentary speech, is its true root. She suggested that for this reason research into vocal sounds for the roots of language function in apes, was probably misdirected. Subsequent events, such as the detailed observation of symbolic behaviour in chimpanzees, the sophistication established in sign languages by the deaf and the development of new computer languages, lend support to her early contention that vocal sound was not a necessary condition for language.

Much symbolic activity must have occurred in association with hunting strategies, socialization, and technology during the 1,500,000 years of hominid life. It was over this period that the hominid brain developed from a volume of 500ml to that of 1,400ml with the emergence of human beings with skills similar to our own. In broad terms, it can be surmized that brain growth occurred during this period, which included the first use of the technologies associated with fire, stone, wood, hides and skins. Since then, the materials we have used have been a basis for symbolic activity, have generated language and provided the means for modelling abstract ideas.

In order to appreciate the importance of their influence upon perceptions and lifestyles, it would be necessary to review what human ingenuity has achieved with each available material. This would be an impossible task and in Chapter 6 we shall consider the value of thinking of materials as belonging to groups or categories. All materials are influential, so by selecting just one category here we make what is virtually an arbitrary or random choice.

However, for the purposes of discussion we will select the group of materials from which we make all kinds of 'sheets'. These materials were available before the outset of human life and since then they have been developed, refined and put to many uses and have provided the basis for many technologies and for the construction of many concepts.

THE BEGINNINGS OF SHEET TECHNOLOGY

Hides and skins have evidently been important throughout history for they were amongst the first available sheets of durable material. Scrapers for cleaning skins were amongst the first tools made and there is evidence of the use of bone tools for removing skins from carcasses and then dividing them. No doubt large leaves and some kinds of bark would already have been used, but skins have a much longer life and provide better shelter, insulation and protection.

It may be assumed that concepts like the need for shelter, insulation and protection did not spring unheralded into the virgin mind of homo erectus. Innovation in technology usually occurs when the inadequacies of an existing technology have been contemplated. In this case, protective places and materials like branches, leaves and loose bark provided shelter of a sort and would, when contemplated, give rise to ideas for their improvement. The limitations of small leaves give rise to the need for larger leaves or sheets and for the need for methods of joining. Branches serve to support sheet materials and provide a model, along with many others in nature, for the idea of a supportive framework.

In spite of their usefulness, skins have the severe limitation of coming only from a live source and were available, at first, only near to where they grew. Short-haired skins made wraps or carrier bags, facilitating travel and the exchange of food and tools. Stretched over frames, they could provide shields and shelter, and make containers and small boats. Fur-bearing skins are still used in some cultures as hangings, floor coverings, bedding and clothing and still function, in genuine or simulated form, to identify persons of rank, in ceremonial dress, and to represent luxury and sensuality in display. This is an enduring symbolic function of the kind which is attached to valued materials. However, for more practical functions, skins have been largely replaced by other kinds of sheet materials, the manufacture of which began with spinning and weaving in Neolithic times.

Inventing the rope
MacKinnon writes:

> 'One of the first inventions of man must have been the rope, a pliable twine of plant or animal fibre, that could be used to bind and tie objects together so that they could be carried round the neck and waist or suspended from any natual hook. The uses of the rope were so multiple that it became the corner stone of many future inventions. In less-developed countries twine is still highly valued and even in affluent industrial societies there survive strong taboos against throwing away string: "So useful you never know when you may need it." We hoard our little balls of twine in equally treasured tins only to discover them untouched years later. Finds in Paleolithic deposits tell us that Stone Age man already knew about string and used it to thread small stones and hollow bones together as necklaces and to fasten heavier rocks as bolas weights. Probably, however, rope is much older than even the cave-men, indeed it can be classed as one of the ape tools.' *(1978, p170)*

In support of his suggestion that the technique of twisting fibres into ropes must have been an early one, MacKinnon describes what happened when zoos began to provide straw instead of woodshavings for bedding:

> 'Almost immediately, in several different zoos, orang-utans began a curious pastime of making rope from the new bedding. One young female I watched in London Zoo was especially proficient at the art. Her game was to carry a bundle of bedding up to the roof, twist it into a long sausage, loop this over one of the bars, twist both loose ends together and then launch off like a pendulum on her rope swing.' *(Ibid, p171)*

For humans, early models for finer strings and yarns would arise from thongs cut from skins, and linear materials like animal gut and vegetable fibres. Once spinning and joining by sewing and knotting were established, darning and weaving would follow. These were the first manufactured sheet materials and the foundation, not only of textile technology but of the enormous range of conceptual models which it has made possible and have become woven into our language.

By 3,000 BC the technique had developed in Egypt to the weaving of very fine linen for swathing mummies, using simple looms with hand-operated combs, to raise and lower the warp to pass the shuttle. The technology of papyrus was also developed in Egypt. Here another sheet material, with different properties, was devised to meet a different need. The stem tissues of a reed were cut into strips, steeped in water and laid in layers at right angles and pressed or beaten together before being dried and burnished to a smooth finish. The resultant 'paper' was receptive to ink like the calendered surface of modern paper, but its structure differed from the felted nature of later papers and was more like the rectilinear structure of modern plywood. The contribution of woven textiles and lightweight surfaces for writing and printing to the development of human thought is undeniable.

SOME EFFECTS OF SHEET TECHNOLOGY

The advent of textiles greatly reduced dependence on the availability of animal skins and, by the development of alternative clothing, made possible the adaptation of human life to less hospitable climates throughout the world. Different properties found in animal and vegetable fibres helped to reflect or retain heat and provide a range of qualities of wear.

It is difficult to do justice, in a few short paragraphs, to the immense influence which the advent of textiles has had on all subsequent cultures. Applications for flexible sheet materials are everywhere: from domestic furnishing to aeronautics and space travel. Some applications are traditional: flags and banners have long been used as rallying points for both warlike and peaceful purposes, from formal heraldry to the local village fete. Sailing, too, employed sheet technology. For many centuries, the use of sails was limited to harnessing a favourable wind, and sailing in other directions relied upon the application of muscle-power by rowing. The development of tougher canvas drawn tighter made it possible to tack into the wind. This was the technology which allowed us to harness the energy of the wind and make the great voyages of discovery and trade. Thus, first by providing clothing and second by harnessing the wind, textiles have served in two major ways to extend the habitable environment.

The technology of the flexible sheet, now available in a wide range of materials, derives from textiles but there has been a parallel development in rigid or semi-rigid materials. This technology has used both natural and synthetic fibres and a variety of glues and cements, to make sheet materials which have limited flexibility and great strength. These materials can be seen in architecture, furniture, boat building, aircraft design and all kinds of screens and enclosures.

Impressive though such a contribution to technology has been over the centuries, the contribution to thought and language has been even greater, and our language contains many words and structures drawn from the textile crafts. For example, the very word 'text' which we use to describe a body of written language is derived from the Greek word *texere*, which means to weave.

The wall, the canvas and the sheet of paper have preceded the video and computer screen as surfaces of attention for most of our writing, calculating and depicting of the world. The combined efficiency of the blackboard and the sheet of paper has played a dominant role in setting the style of our education.

In Chapter 6 we shall look closely at some of the other materials which have always been available to our species. We shall see how they have helped us to generate technology and construct ideas which continue to influence our perceptions and lifestyles.

4 KNOWLEDGE, TRAINING AND STANDARDS

WE have argued that present-day design and technology are the outcomes of centuries of human response to materials and their potential. We have suggested also that primary-school teachers make use of that insight in their teaching. However, it is also true that design and technology make important contributions to all aspects of culture, including the competitive, commercial and industrial efforts upon which standards of living and the economic health of a community depend.

One of the responsibilities of a school is to enable children to maintain the technologies upon which a community relies. Every generation, in every society, makes that demand upon its young. Schools are also expected to enable children to benefit progressively from training in skills and to reach standards upon which a community can rely. Success in such an enterprise requires the extension of children's perceptions, to include awareness of how their learning relates to the wider world of work and community. The following consideration of some aspects of knowledge, training and standards in those wider contexts may be useful to teachers.

THE VALUE OF KNOWLEDGE

It is the nature of knowledge that it bestows advantage upon its possessor. However, since information and skill can both be acquired by the diligent, such an advantage is temporary. There is competition both for information and for skill. Knowledge is specific and whether it is based upon information or skill, the value of knowledge can change according to the problem of the hour or the trends of a much longer period. The two kinds of knowledge are different, but they are related in ways which influence our attitudes to training and to the standards which we inherit.

The advantage of having information is limited unless its possessor also knows how to exploit it. For example, the knowledge that certain goods are to be scarce is of limited value unless accompanied by knowledge of how to obtain, or to manufacture, the type of goods required. If not, the advantage, and the power that goes with it, shifts for the time being to those with relevant technical knowledge. Traditionally, those with skills have looked to those with information for their employment, but this appears to be undergoing a change. To the extent that these two kinds of knowledge are held by different people, the economic health of a community may depend as much upon how the holders use their power as upon how they use their knowledge.

MYSTIFICATION

If knowledge confers advantage, ignorance confers some disadvantage. Those who have knowledge of technology can be at ease in its presence, where others may be anxious about it. The field is such an extensive one, however, that most of us can identify aspects of technology with which we feel reasonably confident. Traditionally, those who understood the 'mysteries' of cooking were not expected to understand car maintenance; these stereotyped expectations are now breaking down and have no place in education. However, we need to understand that they came about, at least in part, from the tendency of those with knowledge to 'sell', and to mystify quite straightforward processes. Art, technology and science are all prone to mystification and not all exponents hasten to deny the special powers attributed to them by those willing to buy their wares.

The prospect of being initiated into the mysteries of a closed group has the strong appeal of privilege and has long provided a motive for learning. Those tempted to exploit it in education should consider its limitations. Its tendency to perpetuate stereotypes and maintain divisions may deny knowledge to those who could make best use of it.

Of course, knowledge which is not applied confers little advantage, so there is a need to share knowledge, under safeguards, with those who possess skills of application.

INSTITUTIONAL SAFEGUARDS

There are two main traditions which exist for the safeguarding of knowledge and standards. The first group consists of the professional bodies, the associations of practitioners of all the acknowledged professions, the great city companies, the learned societies and the institutes of various fields of engineering and technology, merchant and craft guilds. The main concern of these institutions has been the conservation of standards in particular fields of knowledge.

The other tradition embraces the academies, colleges, institutes, high schools and institutions of higher education and training, whose main concerns are with the furtherance of knowledge and the development of training. No institution falls entirely or exclusively into one tradition, indeed the universities have broadened their teaching and training to include new areas of knowledge as they have come to recognize them.

In the first tradition, safeguarding was achieved by limiting membership of institutions to those deemed qualified to share in maintaining standards and extending knowledge. These institutions have the additional effect of conferring status on their members and it is useful to be mindful of the purposes they serve as well as their limitations.

Institutions may now feel the need to ask whether their inherent exclusiveness may serve to maintain impermeable divisions in our training systems, which run counter to the more pressing purpose of raising standards across the whole spectrum of knowledge and capability.

Most professional and technical institutions are only indirectly involved in training, although they have exerted some influence over training and education through examination systems.

Within the second tradition there is a tendency to dissuade students from specializing too early and to encourage them to keep their options open. This postponement of choice may appear to run counter to the raising of standards, since early commitment to a particular field of study is often the mark of high potential. Perhaps each of these countervailing tendencies has merit but suffers from anxiety that its opposite will damage the high standards which each has fought to protect. Such fears might be dissipated by exploring ways of genuinely increasing the permeability of the divisions between disciplines, particularly for those students who demonstrate the capacity to contribute in more than one field of study.

Dialogue does occur between professional societies and academic boards, but even these negotiations appear to occur within vertical bands of specialist 'subject' interests. It is necessary to ask whether the power exercised by institutions which set standards allows training systems to respond quickly enough to rapid change. For example, technical knowledge in various fields of industrial design was in demand for many years before appropriate education and training systems were set up in the 1960s and 1970s. Some progress has been made since then but the whole process could usefully accelerate.

The country house estate gave employment to many people skilled in the crafts required to maintain the household and its environment. The materials employed externally such as clay, stone, lead and glass, were matched indoors by wood, plaster, linens and yarns and the skills of the kitchen, laundry, cellar and workshop.

Problems passed on to schools

Similarly, technology in schools, as a vital aspect of education for all, is at last a requirement of the National Curriculum. Perhaps its late and recent arrival reflects the slowness of our institutions to adapt to the fact that access to knowledge increasingly depends upon technological expertise. Massive changes in the way that information is generated, stored and retrieved, are now taking place. As always, new perceptions will follow in the wake of new technology but only for those whose education and training enables them to be familiar with it. Changes on the scale that we are experiencing will make exacting demands on the adult world to pass on its new technology to the next generation.

However, in spite of the arguments in favour of practical activity and technology in primary education, and in spite of the urgency of developing a generation of young people capable of handling technology in a responsible way, there will be an expectation that old yardsticks can continue to measure progress. The primary teaching profession will need, increasingly, to assume responsibility for identifying and publicizing its new expectations and standards. To achieve this it may initially need to communicate successfully with some of the institutions which share its concerns.

REUNITING SKILL AND INTELLECT

There is a tendency for society to evaluate and separate skill from intellect, underestimating not only the role of the mind in skill, but also the importance of skills in developing attributes of the mind.

It is necessary for teachers to recognize that skill is an attribute of a whole person and is acquired by study and experience. It consists of a persistent intellectual capacity to identify and solve problems which arise when working with material. It may be characterized by much contemplation, as in landscape gardening or computer programming, or by speedy responses as in throwing a pot or landing an aircraft. It may demand considerable dexterity but it is never entirely a matter of the hand.

It is through sensory involvement with technology that the mind is furnished with the models for constructing and communicating ideas.

The separation of intellect from skill has allowed the tradition of separate subject teaching to dominate the secondary school curriculum. One effect of this is that 'study skills' are taught at sixth-form level as though they were separate from the acquisition of knowledge throughout the school career. Many study skills that feature in primary education may not be adequately represented again until post-16 education. The domination of the school day by the different requirements of discrete subjects means that children may not have the opportunity to practise anything for longer than 70 minutes at a time and often for not more than 35 minutes. From time to time, sustained experience is needed for the development of skills and of the perceptions which are necessary for the recognition and realization of new possibilities.

Developing technique
Unlike skill, technique exists independently from the person. One may say to someone attempting to build a wall: 'There is a technique for doing that' and proceed to teach it. It might then be necessary to say: 'You will find that it requires skill and you will have to practise'. There are many techniques for wall building and they exist as part of the store of knowledge which can be called upon. Any technique makes demands in the form of difficulties and the capacity to overcome them with consistency requires practice and judgement. The application of judgement may be missed by an observer who does not have direct experience of the technique. This can lead to failure to appreciate and reward the technician. It can also lead to a failure to demand and reward work of the best quality. Design education cannot be reduced to the teaching of a series of techniques. However, since the need for techniques arises in the course of practical work the teaching of design can provide opportunities for children to be taught techniques in circumstances which are appropriate to their use.

As we saw in Chapter 3, when we looked at developments in sheet technology, invention often arises out of realizing that the objectives of an existing technology could be better met in another way. The limitations of a first technology provide the conditions for the advent of a second. We experience something of this kind of development in conversation when, struggling to state a point of view, we feel obliged to articulate a familiar idea differently. This experience, of thinking aloud so that a group can improve on its ideas by sharing them, should be encouraged in schools.

LINKS BETWEEN TECHNOLOGY AND THOUGHT

Looking at the ways fire developed into the phenomenon of artifical light is a useful demonstration of the intimacy which exists between technology and the development of thought. The fact that fire produced both heat and light is likely to have been familiar long before it was used to light up a dwelling and there would follow a process of discovery of better fuels which ultimately led to the oil and wick lamp – a more satisfactory and longer-lasting invention.

During the long, slow development of portable lighting, each invention was intended to do the job better than the last: animal fat, beeswax, vegetable oils, pitch, tallow, paraffin, acetylene gas, coal and gas, have all been burned to provide light and are still used in some parts of the world.

The first electric lights made use of the carbon arc, literally a bright spark, which needed attention and adjustment. It was extensively used in lighthouses, search lights and early sunlamps. For convenience and general-purpose efficiency its successor has been the filament bulb, in which the passage of electricity through a thin wire causes it to heat up and emit light. As it does so the wire slowly disintegrates until it fails. The life of the filament is lengthened by the use of an inert gas within the bulb to slow down its disintegration. This method of producing light is comparatively wasteful, in that it generates rather more heat than light. Modern tube lighting makes use of mercury vapour which generates ultra-violet light within the tube, and of phosphorescent material on the walls of the tube, which absorbs the ultra-violet light and emits it as white light. This system is more efficient in generating more light and less heat. It is evident how throughout these developments each technology did not invalidate its predecessors but improved upon some aspect of them.

Further, at each stage of the development of lighting, there have been new questions for scientists to ask regarding: the properties of fuel; the behaviour of oils, gases and flames; the nature of energy, its storage and conversion into heat and light; light and other forms of radiation; the visible and invisible constituents of the spectrum, and so on. Each question has led to new ways of seeing the world and has added new models to our language.

HUMAN FACTORS: We have been looking only at the technology of lighting and some of its challenges to intellect, but every development has offered opportunities for design and for the setting of new standards and expectations. Not all technological improvements are suitable for general application: the less efficient tungsten lamp is still widely used and is preferred in many applications to the strip-light because it is more easily shaded, directed and reflected. In many circumstances the eye benefits from variations in the intensity of illumination. Efficiency in material and technical terms does not always mean efficiency in human terms, for example in matters of safety or in the effect on quality of life or work. Research into the full range of human factors which need attention as we design the environments, systems, tools and lifestyles of the future, is still in its early stages.

Complex technology can distance us from the effects of its applications. The most dramatic example may be drawn from push-button weaponry where the effects of use are not seen by the user and may come to light only well after the event. Even in benign fields like medicine, agriculture and automation, the desire for results has to be tempered by forethought and vigilance in design. Responsibility for the design of the outcomes of technology does not diminish because it becomes more difficult to exercise. Survival continues to depend, as always, upon the right choices being made. Education for capability in design and technology is required to meet that challenge.

DESIGN AND MANUFACTURE

The role of designer as a specialist in making choices is a modern one, since making, inventing and designing were once the work of the same person or group.

The craft tradition
Innovations probably arose then, as now, from familiarity with three factors: the material, the process and the objective being met by the technology of the day. In the small medieval workshops the presence of these factors was symbolized in the roles of the personnel: the learner apprentice was involved with preparing raw material and serving the skilled artisan who knew the work and led the process; and in charge was the 'master-craftsman' as overall planner and setter of standards. Progress through each of these roles was a condition of membership of a craft guild. It provided a framework for securing personal status but also for the encouragement of innovation and the control of quality in materials, processes and product. The English craft guilds, which enshrined these arrangements for five hundred years, began to break down when towns and markets grew larger and as graduation from qualified artisan to master became much more difficult. The interests of masters and artisans began to diverge, with the former being dependent on the latter, as towns and markets expanded. By the fifteenth century, artisans were establishing their own guilds, while the richer masters became dealers and merchants and were no longer directly involved in the manufacturing processes. The Livery companies represented their interests, wealth and authority and were distinct from the small masters who manufactured for them, thus effectively institutionalizing the separation of knowledge from skill.

The refining and development of processes were then in the hands of the craftworkers and remained so until the Industrial Revolution. The design of their products evolved slowly, taking on the tried and tested forms of traditional objects, in which were reflected both the accumulated skills of the makers and the requirements of experienced users. With the further division of labour brought about by industrialization, opportunities for small-workshop dialogues between material, process and purpose and between the roles of craftworker, technician and management, were much reduced. The roles became stratified and largely impermeable. Educational structures reflected that stratification, with the craft-level courses being set lower than technician-level courses.

The new generation
However, the advent of large-volume production and standardization in the present century accelerated the need for a new source of expertise in design and a new location of responsibility for its standards. These were

Materials in design and technology

This terrace of typical Victorian town-houses built in the countryside is faced in wood which is shaped in imitation of stone. The main rooms reflect the aspirations of the new middle class and show the tendency observed in earlier country houses to display new styles and craftwork as an expression of standards and values. A cast-iron frieze, employing the new industrial technology of the period, originally extended along the whole terrace, below the bell-shaped dormer windows.

provided by the new profession of industrial designer. Many of these new professionals had trained in art, crafts, or architecture and felt able to improve the quality of manufactured goods. Engineering design was largely a separate entity, often in house and relatively anonymous. As methods of production are changed by technology, and the range of uses for which products are designed increases, greater knowledge and more penetrating research are called for from the designer. Planning for substantial investment is ever more necessary and advances in materials, production and marketing need to be matched by long-term design strategies. These involve effective communication between the separate branches of knowledge and expertise which industries employ. The designer struggles to bridge the gap between knowing what is required and knowing how to do it.

As technology continues to extend our capacities, many of the protective devices adopted by institutions in the past may become inappropriate and unnecessary. Threats to standards are of a different kind. Team-work is increasingly called for as well as international cooperation. It is imperative, therefore, that the kind of communication which can lead to respect between those who work in different fields is rapidly developed. Stereotyped images need to be replaced by informed perceptions of one's own work and that of others.

There is nothing new in the fact that changes made by one generation affect the lifestyles of future generations. Technological change has been in evidence since the birth of the species, but what is new is the scale of change which is now possible. Because of the growth of knowledge and technology, our culture recognizes the huge increase in its responsibility to future generations. Responsible participation in that task is likely to involve institutional as well as personal adjustments. We have a greater capacity than any previous generation to initiate changes in the ways that we use the material resources bequeathed to us. These include human and other living resources as well as energy and space. The choices which we make over such changes are on a scale which may affect lifestyles for many generations.

The inclusion of attainment targets in Design and Technology for all children in primary and secondary schools, signals recognition that standards in design are a product of a whole culture. Design is not an extra quality which can be added to products by experts; concern for good design must permeate all aspects of education as a demonstrable reality.

Quality remains an elusive abstraction until it is manifest in every element which contributes to the form of artefacts, systems and environments. The designer cannot add quality at the end of the production line. Quality, and what it means in respect of any outcome or product, has to be described and worked for in every detail and at every stage of a development programme. There must be some adherence to rules, but these rules must be amended as ideas are improved upon.

David Thistlewood *(1990, p14)* believes that 'Design is an activity that engages all human faculties.' Design must draw upon a whole culture, and design education is essentially cross-curricular.

5 THE NECESSITY FOR DESIGN

SURVIVAL AND THE NEED FOR TECHNOLOGY

In this chapter we shall attempt to identify and describe a number of ways in which design is necessary for successful human life. The relationship between humans and the earth itself is a unique one. Most cultures have attempted to explain the circumstances under which their ancestors might have gained their apparently favoured status over all other species. Now it is possible to see how technology brought about the enormous extension of human adaptability which provides this sense of privilege.

All species must adapt to the circumstances of their environment or they perish. This harsh law of nature applies no less to the human species but, as Richard Gregory suggests *(1981, p43)*, we have learnt to deal with it uniquely. When our ancestors began to use materials to their advantage a new relationship evolved between the earth and humankind. In our time-scale this process took many millennia, but in the time-scale of earth's history it was the work of a moment. It constituted an event which profoundly changed the terms of our survival.

Once we had begun to exploit the potential of materials to assist our adaptation to the environment, we were no longer directly exposed to the vicissitudes of nature.

TECHNOLOGY AND THE NEED FOR DESIGN

The protection and other advantages which technology provides, is the responsible exercise of choice. Technology gives us the opportunity to choose and modify the environments which we colonize. Technology provides the means to survive but does not guarantee survival. The means must be selected and applied wisely with a full consideration of possible outcomes. We noted in Chapter 1 that the ability to envisage alternatives for action and to choose wisely between them, showed a link between the human capability for design and the survival of the species.

Tools, which we use for extending our reach and for exerting more power and finer control, make demands upon intellect and perception. In addition, their use needs forethought and planning; preparation for risk requires imagination and courage. For a full discussion of the 'workmanship of risk' see David Pye *(1978, pp4-8)*. Although initially technology assisted exploration and survival, it has been upon the considered and selective use of technology that survival has finally depended. Here we can identify the elements of design in terms of certain distinctly human characteristics. They include a capacity for each of the following:

- both wide and highly focused aesthetic perceptions
- awareness of relevant factors in an existing situation
- imagining new situations
- planning alternative actions and envisaging their outcomes
- rehearsing and then taking actions and evaluating their effects.

This 1930s' sports car demonstrates its dependence upon earlier forms of transport for its design. Its coachwork, headlamps and mudguards all descend from the horse carriage. Design in technology rarely breaks entirely new ground because our perceptions of what is possible are linked to what we know.

Listing the capacities in this way enables us to see that these are all basic human abilities and helps us to understand how they equipped the species to survive by means of applying technology. Their continued development will be required for as long as we depend upon technology as our essential means of survival.

As we have explored new environments it has been necessary to redesign our protection, equipment, organization and way of life. Humans very quickly overstepped the limits imposed by naked biology and we should certainly not survive without our capacity to design. As a species we are the product of technology; with each development and innovation, the necessity for design has increased.

One of the immediate consequences of design, in the successful application of technology, was the opportunity for the mind to escape its preoccupation with biological necessities. When the vulnerability of the body is exposed, the mind must be preoccupied with danger; but with the body clothed and protected, and the defences of technology in place, the mind can explore more than basic survival. Contemplation and speculation become possible and give rise in their turn to new perceptions. Gregory

DESIGN AND FREEDOM OF MIND

suggests that what is amazing about the human species is the extent to which it has escaped its origins. This he attributes to the effect upon us of the technology created by tools, at least as important as our biological background, when considering how myths, and philosophical and scientific ideas were conceived *(1981, p43)*. Not only does the successful application of technology generate freedom and time for contemplation but it also actually changes aspects of the environment and encourages new perceptions.

Balancing the demands of activity for immediate survival with the more contemplative speculation required for long-term survival has been a feature of each successful culture and continues to be a matter of concern in modern times. Alex Hawkridge is reported as saying:

> 'The best advice I can give to other companies seeking to emulate our success is to start by dreaming. Dream about all the things that are wrong with the business you are in, about the things that give you or your customers problems. Dream about alternatives. Problems approached from this perspective tend to lead to the reality of solutions. Good design is, after all, somebody's dream pulled into reality. And this approach must go right through the company ... the accountants have to dream too, to see how, for example, an advanced accountancy system can help solve their problems.' *(The Design Council 1990, p25)*

The design of the exterior of this 1980s' limousine appears to be calculated to hide its interior construction and mechanical parts. It is impossible to tell what materials are used in its construction. It owes as much to architecture as to engineering and creates an illusion of freedom from all material concerns.

Anxieties about survival are commonly expressed in economic terms. Obsession with them can be a distraction from the necessity of designing and making desirable goods. Akio Morita said:

> 'When I asked an American money trader, "How far ahead do you plan?" the reply was "10 minutes". A 10-minute profit cycle economy does not permit companies to invest in long-term development .... We Japanese plan and develop our business strategies 10 years ahead. There are few things in the US which Japanese want to buy, but there are lots of things in Japan that Americans want to buy.' *(The Design Council 1990, p3)*

Design is more than problem-solving. The attempt made by some to reduce design to this single aspect, ignores the responsibility of design to think ahead and to see alternative possibilities. Some problems can be foreseen and eliminated before they harden to the point where solution is the only option. Good design practice involves the identification and selection of those problems most likely to require attention.

There should be no conflict between the urgent and immediate perceptions required for survival and the longer-term perceptions of imagination and what Hawkridge calls 'dreaming'. Both are necessary and should play a disciplined, complementary and confident part in design. To the extent that we may have allowed these two kinds of perception to drift apart in our culture we may have limited our ability to design a way of life which others seek to emulate, and to design things which others wish to buy.

The enlightened teaching of design and technology will go some way to restoring to mutual respect and cooperation these two important aspects of our culture.

PERCEPTION AND DESIGN

Design begins with perception; the perception of how things are and how they might be. Our conscious perceptions are normally limited to what is important to us. We do not make an overt response to all that our nervous system tells us. We can profitably ignore much of what we see, hear, smell and touch, while concentrating on what we have chosen to give our mind to. Choice is implicit in adult perceptions though it may not always be conscious or deliberate. Children, on the other hand, are heavily concerned with exploring their sensations in order to discriminate between those which require further attention and thought and those which can be merely monitored.

There are times when concentration on existing priorities is voluntarily relaxed and we are then able to contemplate or review our perception of things and so become more fully aware of them. At these times we may contemplate their function or their form and find that we take pleasure in them or find them irksome or unsatisfactory.

One of the important functions of design is to provide a balance of environments in which humans can concentrate their perceptions on what is important, at a given period of time, without inappropriate distractions of the other senses.

However, J Z Young points out that the human race has changed its way of life so rapidly in recent times that it makes no sense to say what is the best or proper environment.

> 'People are very adaptable and will and do live under extremely unfavourable conditions. It is characteristic of Homo sapiens that he makes his own micro-climate and to some extent controls his immediate environment. Indeed he is having a far greater effect than any other species on the whole surface of the earth. Yet individually and collectively we remain tied to nature, as we become all too well aware whenever we are deprived of food, comfort, or health.'
> *(1971, p359)*

This is a useful reminder that our escape from our biological origins is not altogether secure, that it depends upon the maintenance of well-designed technology. For many of the world's population, even the biological necessities of life may be put at risk by the unbalanced use of resources elsewhere. We are only at the threshold of designing environments which take account of the effects of one environment upon another and of the enormous range of human and other needs throughout the world.

RESPECTING MATERIAL RESOURCES

Our relationship with the earth continues to draw our attention to the proper use of resources. The extent to which our civilization has depended and continues to depend on the characteristics of the earth's materials is incalculable. In addition to their availability for direct use in technology, those materials provided the initial frameworks for the construction of our most abstract ideas. Rudolf Arnheim's contention, repeated throughout his writings, is that all ideas are grounded in sensory experience.

Arnheim maintained that certain views of verbal language, which afforded to it an exclusive importance as a function of mind, were unsound. He wrote:

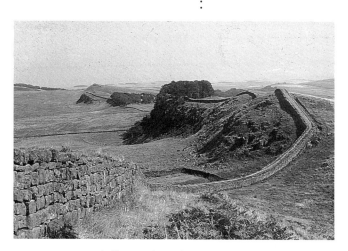

In Roman times the narrowest part of Northern England provided the location and the materials with which to build a wall to mark a defensible northern boundary to Roman occupied territories. The wall has introduced many school children to archaeology.

'Man can confidently rely upon the senses to supply him with the perceptual equivalent of all theoretical notions because they derive from sensory experience in the first place.' *(1970, p233)*

Jerome Bruner suggests more comprehensively that thought is vicarious action and that we work by manipulating representations and models in the mind rather than directly upon the world itself, and so reduce the high cost of error.

'It is the characteristic of human beings and no other species that we can carry out this vicarious action with the aid of a large number of intellectual prosthetic devices, that are so to speak, the tools provided by the culture. Natural language is the prime example, but there are pictorial and diagrammatic conventions as well, theories, myths, modes of reckoning and ordering.' *(1974, p20)*

MODELLING

Modelling, in whatever material, language and modes of reckoning, is the essence of education and design. This is precisely because models allow us to consider, anticipate and rehearse the alternative courses of action we may take. In both education and design, it is evident that we can achieve more by these means than we could by acting directly upon the world. It is, nevertheless, important not to confuse models with reality; we must expect that reality, when encountered, is always rather different from any model, no matter how useful.

The range of vicarious action described by Bruner is no less than the range of thought made available to us through the modelling and metaphoric potential of our encounter with materials. That there are other sources of thought is not in doubt, but perhaps we have seen enough of the importance of matter in the make up of mind, not to leave to chance the range of materials which we make available to children throughout their school education.

From our vantage point in time, near the end of the twentieth century, with a long history of technology to look back upon, we are very conscious of the future – an understandably popular subject for contemporary films and fiction.

Awareness of rapid change within our lifetime forces us to speculate about a future in which we confidently anticipate further change. J Z Young writes:

> 'But man's special feature is the power to make plans which will alter his surroundings. He often does this in ways that allow only the barest survival, in the crudest shacks or the harshest concrete blocks of apartments.' *(1971, p360)*

This is a biologist's view of the characteristically human ability which is to design for survival. Most animals are able to take evasive action when threatened but what seems to be uniquely human is the ability to exercise choice in such action, and to reflect on it to the point of preparing to meet future threats, by actually changing the circumstances in which they will be met. Weapons and defences are not developed in the heat of battle or the hunt, but in contemplating the efficiency of existing ones, for future confrontations. When considering improvements to defences, hollows could be deepened, traps laid, shelters and hides built, strategies developed, and protective shields and clothing made. Planning of this kind involves making informed guesses about the future.

It would be a mistake, however, to think of primitive life as wholly preoccupied with hunting, food gathering, shelter making, defence building or other purely survival activities. Primitive life in favourable circumstances can be comfortable, and a long period of adaptation can reduce survival activity to a minimum. In such periods of relative tranquility, the imagination is able to speculate more widely about the future and the shaping of the environment.

Young continues:

FUTURE-ORIENTED DESIGN

Materials in design and technology

In this photograph a child has painted the head of a tiger on a bedroom window. Children need to see evidence of their effectiveness and worth, in their family or the wider community, throughout their childhood and education. In this case the child was able to take possession of the room and share responsibility for it by decorating it.

'Yet since Paleolithic times he has decorated his homes and meeting-places, and loves to do so still. There is some connection that we do not understand between our powers to think symbolically and the urge to have symbolic structures around us. From very early history there are signs that man decorated the artefacts that they made .... It is clear that the creation of beautiful and symbolic objects is a characteristic feature of the human way of life. They are as necessary to us as food and sex. But some form of symbolism and decoration are features of all societies. Not only do we individually need and want the arts, but as a group we neglect them at our peril.'
(Ibid, p360)

Design has the responsibility to impart to the environment, and to all the objects which furnish it, a true reflection of the values of the time. The best of design, at any period, reaches beyond styling and the updates of fashion to provide common access to the highest prevailing values and aesthetic sensibilities. The nobility of the form of a great Gothic cathedral or of a Doric temple is legible and speaks of the spiritual aspirations and awareness of the cultures which produced them.

It is important to acknowledge the use which is universally made of activity leading to symbolic enrichment. It is customary to say of such activity that its purpose is to nourish the human spirit, and that claim is certainly made of artistic activity generally. However, it must be argued that the kind of contemplative activity which reaches beyond basic necessities is itself essential to the development of technology and therefore to the survival of the human species.

Bruner makes what he calls the rather unorthodox suggestion that:

'In order for tool using to develop, it was essential to have a long period of optional, pressure-free opportunity for combinatorial activity. By its very nature, tool using (or the incorporation of objects into skilled activity) required a chance to achieve the kind of wide variation upon which selection could operate .... Dolhinow and Bishop (1970 p142) made the point most directly. Commenting first that "many special skills and behaviours important in the life of the individual are developed and practised in playful activity long before they are used in adult life", they then note that play "occurs only in an atmosphere of familiarity, emotional reassurance and lack of tension or danger".' *(1972, p38)*

Design for survival requires focused perception and pressure-free contemplation, associated here with a necessary 'playfulness'. Teachers should be encouraged to recognize the survival imperative which attaches to contemplative studies which may not, on the face of it, appear to be utilitarian. They need to continue to educate themselves, parents and children of the importance of activity with materials as a way of generating ideas, which are capable of being used in real life.

The biologist, Young, summarizes:

'If we are to ensure a satisfactory survival we need to pay deep attention to the provision of an environment that is adequate both in its architectural design and symbolism in relation to nature. There is a danger in a self-propagating technology that takes over the whole earth without considering either the men or the other creatures on it. The arts can provide our defence against this if we allow them to do so. They have the most central of all biological functions – of insisting that life be worthwhile, which, after all is the final guarantee of its existence.' *(Ibid, p360)*

While the diagnosis of danger given here is sound, the safeguarding role which Young rather vaguely gives to the arts is over-generous and may itself constitute the danger of allowing technologists or others involved in design and decision making, to believe that they can choose 'bare-survival' solutions on the grounds that the arts will in some way be able to retrospectively redress the balance. To see the arts generally as remedial to the worst excesses of technology is to fail to understand their relationship: the arts are a necessary factor in the progress of technology.

It is necessary to emphasize that design has always been necessary for personal and cultural survival and this has been achieved by providing guidance for technology and obtaining in return some leisure for contemplation and new perceptions. These functions for design are still necessary and to them is now added the function of improving our perceptions of human and ecological well-being.

6 MATERIALS AND IDEAS

CATEGORIZING MATERIALS

We chose in Chapter 1 to use the word 'materials' to mean any physical matter or substance which has been found to be useful. It is necessary in this chapter to propose some simple groupings of materials in order to consider them in greater detail. There are a number of ways to approach this categorizing task.

Scientists, for example, now know a great deal about the molecular structures and behaviours of elements and compounds and use a number of categories based on that knowledge. Their use of terms like: metallic and non-metallic, organic and inorganic, pyramidal or planar, allows them to describe structures and to anticipate behaviours.

Technologists tend to refer to materials by their function, with words like: abrasives, aggregates, fluxes and fuels. Craftworkers may use categories that reflect properties which are of importance in their craft with words like; plain or figured, natural or synthetic. Artists' materials are often grouped according to their usefulness as media, hence: oils, acrylics, collage, film, papier mâché and so on.

Each discipline acquires a vocabulary, and sets up categories according to its methods and practice.

In primary education, it may be helpful to think of materials available to our ancestors from the beginning, before the disciplines as we now know them, were established. Since there is, however, no such thing as a 'beginning', any effort to list such a range of materials is an exercise of imagination. The list below is a typical outcome of such an exercise carried out by a group of teachers.

The list is headed 'aboriginal', which means 'from the beginning' and for Australian aborigines at least, the mythological beginning is lost in that pre-history which they call 'dream time'. Although the list is mythical it may, nevertheless, be useful. It helps us to realize that it was from a range of materials no more extensive than this that any one human group or settlement began its progress toward what we know as civilization. Apart from the technology and intelligence bequeathed to them by earlier species and the examples of structures in plants and other creatures, there was little to guide them. All of the technological and other cultural achievements which have given form to our civilization have grown from human response to such a range of found material.

As a starting point, here is a list of useful materials available from our 'mythical' beginning:

ABORIGINAL MATERIALS

- air, antlers, bladders, bones, boulders, branches, blood, caves, chalk, charcoal, clay, creepers, dung, earth, feathers, fibres, fire, furs, gases, gourds, grasses, gum, gut, hair, hollows, horn, ice, juices, leaves, logs, mud, oils, pebbles, quills, reeds, saliva, sand, sap, seeds, shells, skins, snow, stems, stones, teeth, thorns, trees, tusks, twigs, water, wax.

There is nothing authoritative about this list; in fact, one of its strengths is that it can be reconstructed by any group of people who might wish to consider the material foundations of technology or the influence which technology has had upon ways that we think. Having constructed such a list, it is then possible to attend to each material in turn.

However, it may be more convenient to group them according to their most obviously useful characteristics, in order to look at how materials generate and embody ideas. Some materials, for example bone, appear in more than one group because of the many different forms in which they can be found.

This honeycomb is just one natural structure imitated by designers for strength, lightness and insulation.

In each category, below, we look at the basic uses made of materials. Some uses exploit the characteristic form of a material; such as the sheet-like nature of the leaf or the rod-like nature of a branch. Others exploit the properties of their substance; like the intensity of a dye or the adhesiveness of clay or the gum of a particular plant.

The remainder of this chapter is a consideration of each of these groups

Materials in design and technology

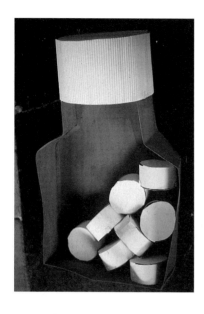

of materials in turn. Its purpose is not to provide an exhaustive reference but to give some idea of the ways in which materials generate ideas and of the enormous provocation to learning and understanding which activity with material can provide.

- Rods and levers: branches, long bones, logs, stems, tusks.
- Gases and fluids: air, water, oils, saliva, sap, juices, blood.
- Dyes and pigments: chalk, charcoal, earth, blood, juices.
- Tools, instruments and small units: antlers, quills, stones, teeth, thorns, twigs, pebbles, sand, small bones, seeds, shells.
- Strings: creepers, hair, feathers, fibre, fur, grasses, gut, reeds, stems.
- Sheets: leaves, skins, ice.
- Plastics and adhesives: clay, dung, gum, mud, snow.
- Containers: bladders, gourds, horn, hollow stems, shells.
- Large units: boulders, caves, hollows, trees.
- Food: all animal, vegetable and some mineral matter.

RODS AND LEVERS

Rods and levers were amongst the first 'found' or 'ready-made' tools to be used to extend both the reach and the efficiency of humans. A more or less flexible stick in the hand extends the reach for collecting food or to reach over a fire to cook it. Simple though the idea is, it remains the first example of 'remote control'. We now use a much more refined version of this practice to perform operations upon the arteries of the heart, making entry through an artery in the thigh, without having to resort to invasive surgery through the chest.

Levers, when used as hand tools, can greatly multiply the effect of human effort. Two kinds of lever work by converting long movements into stronger short ones. The first of these is like the tyre lever, where a long downward movement of the hand is converted to a strong upward movement of the tyre wall as the lever rocks around the fixed edge of the wheel rim. In this kind of lever the fixed point or fulcrum is between the hand which applies the force, and the load to be lifted or the work to be done. In scissors or pliers, two levers of this kind share the same fulcrum, and do the work between them.

The second kind of lever works as in the typical nutcracker, or foot-pump, where the fulcrum is at one end with the hand at the other. The work is carried out at a point between the hand, or foot, and the fulcrum. Again, because the lever rotates about a fixed point, the end where the hand is applied travels further than the point where the nut is cracked, so converting the effort into a shorter, stronger one.

There is another kind of lever, like the eyebrow tweezer or sugar tongs. Here, the fulcrum is again at one end but the effort is made between the fulcrum and the place of work, which is at the opposite end. In this case, a

Materials and ideas

short-moving effort is converted into a longer, though less strong, movement, which may be more useful; for example in waving a distress flag at the top of a long pole. The human forearm mainly functions as a lever of this kind; the act of raising a cup to the mouth requires a short movement of muscles in the upper arm but they have to exert a force some six times greater than the weight of the cup to achieve it.

The internal pulling strength of human muscle, measured directly, is enormous. Its value lies in the fact that the length of our limbs can convert it into both range of movement and speed of operation where it matters. The greater length and articulation of our limbs gives us superior control when compared with some other mammals; they can exert superior strength in defence or attack but with much less versatility than we possess. Our relative weakness, which contributed to our need for technology, was the price we had paid long before, as a species, for the development of the articulate limbs and minds which enabled us to continue to develop technology. The conversion of energy from one form to another for its more effective use is a continuing challenge for technologists, designers and ecologists.

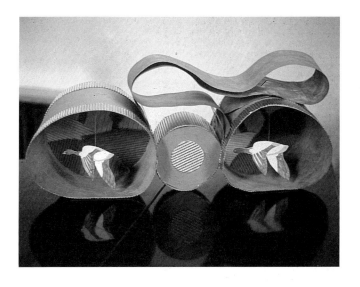

A few modern examples are used here to describe the different types of lever but, in practice, hand levers are in use all the time. Where one hand applies the force the other may act as fulcrum, and the roles can pass from one hand to the other, as in digging and shovelling, or from one finger to the other as in sewing, knitting and even in writing and drawing. The body acquires familiarity with all three kinds of lever without their mechanical description being expressed verbally. For example in casting, by rod and line, or hauling in a fish by the same means, the lower hand acts as a fulcrum and is held fairly still while the upper hand, between the fulcrum and the load, has to work hard to produce a large movement at the top of the rod, even when the load is quite small. In actions such as sweeping, raking, hoeing and scything, the long handles provide leverage which allows for greater achievement.

Kinaesthetic knowledge of leverage begins, as we saw in reference to the baby in Chapter 1, in the way the body relates to the world around it. Children at three or four years of age go on to demonstrate considerable agility and, if they experience the scope and provocation of an environment which promotes climbing, learn how to exploit their favourable power to weight ratio.

The models on these pages all include cylindrical forms. A class of junior children were shown how to form a closed or open-ended cylinder from thin card and were asked to construct a model of a familiar object. Opposite: an aspirin bottle; top: a pair of binoculars; bottom: a light bulb.

Materials in design and technology

This 'house' was built by an 8-year-old girl who sought a sheet of corrugated iron and a bag of cement for her birthday. We must not assume that play follows role stereotypes. In our selection of the materials we provide, we can take steps to minimize this tendency and counteract any tendency in children for their responses to become merely habitual.

Understanding the rotation of levers about their fulcrum provides the key to the understanding of the whole of mechanics. The wheel extends the function of the lever through a complete circle. There are no wheels as such in nature but it is, however, implicit wherever anything moves about a fixed point, as in swinging, dangling or rolling.

The following concepts and linguistic ideas involve models drawn from the experience of rods, levers and the mechanics which derive from them.

- Balance: the seesaw; the beam balance; scales of justice; symmetrical and asymmetrical balance; equality; the equation; proportions and fair shares.
- Structure: the beam or lintel; a crosspiece supported by two upright posts; the basic doorway, an essential architectural and structural concept as found in the bar, the step, the ladder and the stair; the pole; the shaft which, when stressed, becomes the bow or spring or when lashed together, makes units for roofing, fencing, stilts and floors.
- Leverage: axe; pick; hammer; club; bow; spindle; arrow; spear; spike; crowbar; drumstick.
- Action: handling; chopping; pickaxing; polaxing; hammering; beating; aiming; flighting; throwing; spiking; levering; weighing.
- Symbols: staff or rod of office; pointer; cane; cross; flagpole.
- The wheel: rim; spoke; hub; cog; pulley; revolution; rotation; reverse; ring.
- Metaphors: taking the wheel (to steer); putting one's spoke in; big wheel; inner wheel; put shoulder to; wheels within wheels; break on the wheel; oil the wheels; spin the wheel; apply the brakes.

GASES AND FLUIDS

The most significant early use of materials in this category was of wind and water as sources of energy. The earliest technologies were simply an extension of human muscle-power. For example, windmills and watermills were the major sources of power, other than humans and yoked animals, until the invention of functional steam engines in the seventeenth century.

Air and water are unique in being even more essential to us than food. An intake of air is required every few seconds, and water every few hours. Air and water also function as media through which we travel. Our relationship with these two materials is especially intimate at a chemical, cellular level and in the pneumatic and hydraulic functions of our bodies. Acknowledgement of this special relationship, if not of its chemical nature, is reflected in mythology, literature, and religion alike. The myths and literature of every culture personify or deify water and air.

We use water for bathing and cleansing (both actually and symbolically) and other fluids internally and externally as medicines and unguents.

Materials and ideas

We use hydraulic and pneumatic systems for the storage and distribution of energy without which our transport systems would grind to a halt. For one hundred and fifty years, until the invention of the internal combustion engine, external combustion of coal produced compressible steam to bring forward the achievements of the Industrial Revolution. This century has seen its replacement by petrol, diesel and electrically powered engines. The price of fuel oil and its sensitivity to the vagaries of supply and rumours of shortage is a reminder of the importance we attach to this mineral.

We have made use of fire for at least five hundred thousand years as a source of light and warmth. Once tamed, it materially changed the scope of human life through technology, rendering food tender and smelting ore into metals. Metal smelting began with copper and bronze about 3,600 BC and iron about 1,000 BC.

Aluminium, although the most common of all metals, was not successfully isolated until early in the nineteenth century, and not highly used until the twentieth century, when its lightness was in demand by the aircraft industry. Very suitable for extrusion, (shaping by forcing through a hole when in a semi-plastic state, like toothpaste from a tube or clay from a pug mill) aluminium makes continuous lengths of any required section.

Iron when first smelted contains many impurities, which render it very hard and brittle. It can then be remelted and poured into moulds to make a pre-determined shape. This imparts to it great strength under compression. Cast iron is widely used where heavy support is required, whereas wrought iron is hammered and rolled into shape and is more resilient. It was not until the mid-nineteenth century that a reliable industrial method of producing steel was developed. Steel, although an alloy of iron and carbon, actually contains less carbon than crude or pig iron. Less than three per cent of carbon makes mild steel a tough, versatile material of great tensile strength. Less than 1.5 per cent carbon produces a steel capable of being hardened to the point where it can be used for making machine tools, which are cut into ordinary steels.

The following actions and ideas also function metaphorically, by providing conceptual models drawn from the experience of gases and fluids:

- thirsting, quenching, soaking, saturating, flowing, streaming, absorbing, washing, annointing, wetting, sprinkling, floating, evaporating, dissolving, irrigating, lubricating, tides, waves, wave motion, pressure, draught, volume.

As Arnheim pointed out, we can find words for our most abstract ideas and so we have 'thirst after truth', 'soak up knowledge', 'absorb information', 'flow with the stream', and 'fail under pressure', because such ideas derive from sensory experience in the first place.

An item made by a child is uniquely their own possession. In this photograph we may see not only pride in achievement but also pride of ownership. A model of a living creature can be the object of love and hate, of care and neglect, but is liberated from the normal consequences of such emotions. This is a key to the value of play as rehearsal for living, which we may make use of in education.

Materials in design and technology

DYES AND PIGMENTS

Body painting, for ceremony and war, seems to have been universal and many of the earth pigments first used like chalk, red iron oxide, ochres and carbon still provide cheap and widely available colouring matter, now mostly used in opaque paints. Adornment of the body was superseded by decorative dress, as human settlements moved into cooler regions, although it is still practised in areas where little clothing is worn. Tattooing and facial make-up survive, with the latter now relating closely to fashion and dress. Make-up is used to highlight natural features or to create a new mask or mood. Masks relate to ritual and theatre and have been much used, in play, as a way of experiencing emotion in an alter ego. Twentieth-century make-up was greatly influenced by the limitations of black and white films where legibility of expression was paramount and high contrast of feature to face was required.

Colour is used in nature to convey signals. It contributes to mood and is therefore important in design. The effect of colour is changed by context and this has to be anticipated in design. In order to obtain full value from colour in design it should be given as much consideration as any other factor in relation to the purpose envisaged. Colour is rarely neutral and is often a very powerful element in achieving a design objective.

Dyes from insects, squid, and vegetable sources were used as inks or for staining white fillers like chalk. Dyes and pigments were also thinned in oils, gums or water to make opaque or translucent paint or printing material. Pictures preceded writing but it is significant that the two means of communicating and thinking have, for the most part, shared the use of pigment and complemented one another.

We commonly talk of the ability to picture things in the mind but any discussion of the process quickly leads to the conclusion that such pictures are very incomplete – at least as incomplete as our perceptions. The pictures in our 'mind's eye' are not confined to visual information received via the eye alone but are composed from all our knowledge and sense data. Philip Rawson writes:

> 'Drawing embodies a genuine and independent way of thinking. Someone who draws actually sees more and knows more of the world than someone who does not draw.' *(1983, p7)*

This simply means that a person who draws learns to structure his or her perception in terms of the drawing medium as well as, like most of us, in words. When it comes to picturing, for example a horse, the person who draws can picture the animal in graphic terms and can, if necessary, draw the picture. Others are much more likely to structure their mental 'picture' of the horse in words, possibly with greater detail and equestrian knowledge. Such a person cannot, however, draw their picture, nor should they

These ropes were offered as starting points for a first-school activity. One child saw the ropes laid together as being like a river and began to tiptoe across them. Soon he had developed rules for a stepping game and began to play it, devising more elaborate rules as required. He was then able to teach the game to his friends. This development was a significant experience from which to learn how to turn the limitations of a given material to advantage.

expect to, since it has not been composed in visual terms, nor with reference to a visual medium. The capacity to draw begins, like design, with perception, it involves learning to see in the terms provided by dyes and pigments. Hence their inestimable importance in our history.

Once made accessible, through drawing and painting, many ideas are then absorbed into verbal language. Similarly, literary ideas can often be translated into visual images by people who are both literate and 'visuate'. Notables such as William Blake as illustrator of his own writing, Sir John Tenniel and George Cruikshank, illustrators respectively of Lewis Carroll and Charles Dickens, have imposed their perceptions as a dominant influence on all subsequent interpretations.

TOOLS, INSTRUMENTS AND SMALL UNITS

Perhaps the first tools after hand-held flints were the pick and axe. Teams working with picks could make the first large excavations, improving fertility and extracting useful minerals. The design for these tools may have derived from the useful shape presented by pieces of antler which were reinforced with flint at the striking edge, a process which involved a taper or wedge fit. These methods of attaching working and load-bearing parts to shafts are still used in precision engineering.

Certain materials were formed in ways which provoked their use as instruments. Pieces of antler were cut and shaped with stone gravers and scrapers to make arrow heads and needles. Thorns, fishbones and quills might have served as tooth picks, and then as pins, probes, nibs, and larger applications such as toggles and buttons. The toggle is a small symmetrical piece, like a single finger bone or fish vertebra, which is used on a rope or rod, to limit its movement through a hole. They are often hinged in the manner of a buckle to allow passage one way but not another; like the expanding device we use to screw fittings to hollow walls. The term is now applied to switches or keys which switch continuous functions on or off.

Small units give us active access to concepts of number, groups and sets which feature in primary education. In primitive cultures, in the form of pebbles, shells, small bones, quills, seeds and teeth, for example, they are used for decoration and personal adornment, and include amulets, and badges of rank and status.

They provide a means of symbolizing, which is a vital element in both language and computation. Small units can be used to represent such entities as people, houses or planets and to reveal relationships and patterns between them. Hence their use in play, fortune telling and games like hopscotch, knucklebones and fivestones, versions of which are universal.

Stones used for building could be simulated by bricks of dung or earth. Such unitary structures extend into forms of building characterized by the shapes of the units and the way they fit together.

The hand quickly benefited from the assistance of small instruments. Perhaps the first of these, the sewing needle, about one millimeter in diameter complete with an eye, was made some nine thousand years ago. Such needles were made by scraping parallel grooves on bone or antler and gradually undercutting the tiny ridge formed between them.

STRINGS: *Characteristics*
Natural 'strings' like gut and fibres were quickly imitated in twisted cord and cut hide. Amongst the properties of different types of 'string' which give them their value are:

- length, consistency, continuity, flexibility, tensile strength.

These lend themselves to the following processes which are involved in a vast range of applications:

- binding, wrapping, lashing, knotting, tying, measuring, leading, tethering, roping, stringing (musical instruments), spinning, weaving, whipping, spiralling, suspending, threading, measuring, taping.

Amongst the technologies involved are the important ones associated with:

- textiles and dress, knitting and weaving, ship rigging and sailing, animal husbandry, transport, surveying and exploring, surgery.

Metaphorical uses of these functions are widespread both in ancient myths and modern language:

- The Gordian Knot: the legend in which a rope made from bark fibre was tied so ingeniously that no one could untie it until Alexander cut it with his sword.
- The Labyrinth of the Minotaur: where the escape of Theseus from the labyrinth was achieved by the use of thread to guide his return.
- Arachne the tapestry weaver: whose skill with her yarns so unnerved Minerva the goddess of wisdom it made her turn Arachne into a spider.
- The Sword of Damocles: which was suspended over him by a single hair as a consequence of his envy.
- Aesop's fable 'The Lion and the Mouse': in which the mighty lion imprisoned by a net of rope is rescued by the sharp teeth and intelligence of a diminutive mouse.

In computing we refer to a set of linked characters or numbers to be processed as a unit, as a 'string', and a self-repeating section of a program, as a 'loop'. More familiarly we can 'string a few words together', 'tie up a few loose ends', 'run a tape over this', 'seem highly strung', 'be tied up for a while', 'gird the loins', and so on.

SHEETS

In Chapter 3 we looked at the development of sheet materials and technology, from animal skins and leaves through textiles and paper. These two technologies alone have provided material support for the art and literature which make up so much of our culture.

Given sheets, we had the maps, sails and scrolls to aid and record our explorations; wall hangings and floor coverings to soften our environments; dress, fashion, costume, theatre, religious, civil and military ceremony, heraldry and flags; as well as the means to harness wind and ultimately to fly.

The woven sheet also gave us tangible access to the idea that areas can be constant though flexible. Paper and card make tangible the idea of the planes and intersections.

Curtains were the first non-weight-bearing walls and have now been complemented by many kinds of sheet materials, such as glass, paper, plastic and compound boards. Metal and synthetic sheets are light and strong and can be further strengthened by bending, fluting, pressing, corrugating and layering. Joining methods include sewing, stapling, folding, riveting, crimping, spot-welding and taping, as well as gluing and bonding.

Design and manufacture in card materials provide a common basis for the packing and packaging industries. Sheet materials can now be made in opaque, translucent, and transparent plastics in a vast range of colours and having properties designed to meet particular purposes. They can be vacuum-formed into lightweight shapes to pack products from trays of chocolate to precious and delicate objects, or in manufacturing objects from instrument boxes to cockpit covers.

In making this model, a child has created a successful visual equivalent of a bicycle in cardboard. In doing so, they had to overcome the technological problems associated with construction in cardboard and not those of construction in steel.

PLASTICS AND ADHESIVES

Plastic materials are those which deform or change shape under stress and tend to remain in the altered shape when the stress is removed, unlike an elastic material which returns to its original form or shape.

In practice the term is used to describe any material which will conveniently take up the shape of a mould. Clay is the most commonly available natural plastic. Modern plastics have been synthesized so as to be pliable when warm, and rigid when cold. Thermosetting plastics do not soften when reheated but thermoplastic ones do. Many materials, including for example iron, glass, and some sugars, are plastic while within a certain temperature range.

Some plastics are adhesive and these include natural gums and resins. Adhesion depends upon intimacy of contact between the molecules of the substances being joined and the molecules of the adhesive material. Plasticity assists this contact, but may cause the joint to be too flexible for some purposes. Adhesives which harden after intimate contact may be stronger. Some adhesives, like natural hoof-and-horn glue, or synthetic

Materials in design and technology

Real bricks require care and effort in handling and lead naturally to team work. Here, the possibilities for using the bricks included the basic post-and-lintel structure which allows visual and actual passage through the walls. Experience becomes particularly valuable where a number of life-sized materials are explored, and their limitations and advantages are compared.

thermosetting plastics, are applied hot and harden on cooling. Others like size, casein or PVA are dispersed in a solvent and change their character or state as the solvent evaporates. It helps if the adhesive responds to heat, moisture and movement in much the same way as the material or materials being joined.

In this respect the ideal adhesive would be the same as the materials to be joined; this is achieved in welding. Two pieces of the same material can be welded when melted by the application of heat at the site of the joint. When the materials begin to liquify and flow together the heat is withdrawn causing the near-fluid material to solidify. A small amount of additional material may be melted into the joint at the same time to ensure continuity of structure across it. The skilled welder has to apply sufficient heat to the joint to cause the material to flow enough to secure a molecular mix but not enough to burn or melt adjacent material to the point of weakness.

Soldering and brazing are similar processes, in that local heat is applied, but the jointing material is either a lead-tin alloy called solder, or brass, a copper-zinc alloy. These processes require less local heat than welding. They are useful where the high temperatures required for welding might weaken or distort the original material.

The joining of clay involves similar principles. When the clay is plastic, it may entail smearing clay across the joint, often in both directions. This ensures continuity of the material, as in welding. Where clay has dried to a leather-hard condition this action is not possible and the method of ensuring continuity involves scoring the surfaces to be joined and adding slip – clay thinned with water – which like solder makes intimate contact, before drying and hardening. Many foods undergo changes of state during cooking, including the temporary plastic state as in dough, and pastry. Many of the processes in baking closely parallel those in pottery and share the same vocabulary.

Using plastic material gives experience of processes such as moulding, casting, extruding, coiling, pressing and rolling; while experience of adhesives relate to adhesion, adherence, tenacity and tenaciousness.

Concepts like loyalty are, of course, possible without experience of plastic materials, but their availability permits such a concept to be 'held' in terms of a range of tangible models which enrich both language and understanding. Thus one may 'stick to one's principles' or 'be manipulated', 'adhere to a cause' or reveal 'feet of clay'.

CONTAINERS

Perhaps the most important early invention of humans was the 'carrier' bag, for it made possible the first transport of a quantity of goods by the first entrepreneur to venture into trade. The handbag and pocket are essential items and the American brown bag and the European plastic bag make supermarket shopping possible.

Natural containers like eggs, gourds and shells typically contain food and lend themselves to its storage. Bladders and other viscera can contain fluids or foods. Natural containers were copied in clay and sheet materials. Once established, the container takes on the potential for possessions, secrecy and concealment.

Containers are useful in many contexts. Grouping 'objects' into sets and categories allows us to deal with them much more readily than handling every idea individually. Ideas, attitudes and observations can be bundled conveniently into containers, and filed for reference.

Aspects of language, particularly of vocabulary, work to this model. One function of a word is to act as a container for a range of ideas and experiences: the category 'dog' may contain a number of small 'creatures' some of which may later need to be transferred to other containers like 'cat' or 'toy'. The process is not confined to childhood; learning may be seen as extending both the range and content of these containers. Accurate classifying requires discrimination and study. Many containers, like the original carrier bag, soon require sub-sections. Two rules emerge from experience: one is the importance of remembering where one 'put' things, and the other is never to mistake the container for its contents.

Classifying and sorting are at the heart of information storage and retrieval for which the computer is so useful. 'Input', 'processing' and 'output' constitute a new use for the container metaphor.

LARGE UNITS

Large natural phenomena might include a large tree, an outcrop of rock, a hollow in the ground, a river bank, a hill top, a natural amphitheatre or a plateau. Human beings seem to have a natural sense of place and most of us, including children, have favourite places where we choose to be at certain times. Pilgrimages are made to places of special significance and geographical features have a strong influence on memory.

Some places or large objects become the location for people to gather in natural informal assembly. These natural features provide the beginnings for architecture and design for community life. Some of these features are built into well-designed cities, towns and villages, but they are still sought in 'natural' surroundings, not only for the amenities they offer but also to meet a particular mood or the need to be 'at one with nature'.

Features of landscape, seascape and townscape give us our sense of scale. They represent to us models of time, space and energy to which we can relate directly and from which we can draw a sense of our own reality. They provide a measure for our responses to what we make and do.

Monoliths

Monoliths are amongst the most simple and compelling of all works of design. All that an observer can see, in many cases, is that a large stone has been rotated until its long axis stands vertical. The concept is simple but the effort is considerable, requiring determination and teamwork. An erect stone in an open space is therefore about as simple a memorial to civilized community as can be deliberately made.

Sculpture as an art form plays upon our feeling for hollows and boulders and it is no accident that sculpture has long been associated with the partially enclosed cultivated space of gardens. We are ambivalent about space; sometimes we seek open spaces, while at others we welcome enclosure. Well-designed spaces provide variety and choice accompanied by grades and contrasts in light and shade. Attitudes to the use of space vary between cultures and between individuals and may be reflected in different ways of moving in structured spaces and observed in traditional forms of drama, dance, and ceremonies.

Humans have a special affinity with trees – another relevant large structure. As children we climb trees and take shelter under them. Our language relates us directly to their trunks and limbs and they provide us with food, fuel, structural material, medicines and, in the form of forests, with the oxygen we breathe.

FOOD Food comes in all shapes and sizes and is the most ephemeral of materials. Some materials associated with food, such as undigested seeds and bones, together with evidence of food preparation and cooking, and studies of dentition, have allowed archaeologists to learn about the diet of early humans. Not all animal or vegetable material is edible and the category 'food' carries the implication of another category labelled 'poison'. The distinction is of such importance is to be associated with the origins of concepts of 'good' and 'bad' and the beginnings of 'moral' education and discrimination.

Food is a special category because it reminds us uniquely that the very structure of our being is a part of the universal continuum of matter.

Much of our time and ingenuity is devoted to ensuring an adequate food supply, through the technologies of agriculture, and the storage, distribution, preparation, consumption and recycling of food energy.

Our spiritual concepts are modelled upon metaphors of maternal nutrition, paternal shepherding, pasture and seasonal maturity. Food and drink symbolize the meeting of the spiritual and material aspects of existence, and provide the basis for social ritual and assembly in most cultures. They provide motivation for all kinds of activity from bare survival to the highest refinements of design and technology. Health and nutrition continue to stimulate speculation and research, and rank amongst the most important concerns for design and technology.

SUMMARY

One purpose of this somewhat arbitrary list of materials, and ideas arising from them is certainly to indicate the important role of materials in directing our thought and providing the means for constructing our civilization and our technology.

However, a second and more important purpose here is to indicate that any individual or group, like teachers or parents, can draw up their own categories of materials and use them to highlight their existing knowledge of materials and technology. As users of technology, most adults have much more knowledge of the subject than they initially believe. This discovery can free teachers from dependence upon those 'survival kits' and packages which are sometimes produced to exploit their insecurity. Provided that it is fully articulated, shared and coordinated, the existing expertise in any one primary school is likely to offer an adequate base upon which to introduce the teaching of design and technology.

If this confident and independent base can then be augmented by growing awareness of applications of the subject, the range and quality of teaching and learning can be extended to achieve rewarding outcomes.

7 NEW FACTORS AFFECTING DESIGN

AESTHETIC RESPONSES

BEFORE considering new factors affecting design it may be useful to review some responses to design and then some criteria used in evaluating it.

We have suggested that design begins with perception. It is tempting to say that design also ends with perception but there is a sense in which design does not end. The outcome of design activity, whether object, system, or environment, is always a temporary one which can be improved upon in the light of further experience. It may be more accurate to say that each stage ends with perception. The designer may seek the satisfaction that each outcome meets the requirements of the initial proposal. Sound judgements include awareness of all the elements which make up the designed object.

The perceptions of the user, client or purchaser of a product may differ considerably from those of the designer. There appear to be two main kinds of aesthetic experience which users can derive from well-designed objects. In the first, the senses are surprised by what they encounter. A staircase or a bridge might provoke comments like: 'I did not know a staircase could be like that!' or 'How could anyone think of such an elegant shape for a bridge?'. It is the kind of experience that adds to one's sense of what a staircase or bridge might be. It sets a personal standard by which all subsequent staircases or bridges may be judged. What matters here is not that the object is actually new but that the observer or user is made freshly aware of it. Sensing in a cave painting the deliberation of its designer, would serve as an example, or passing under the Pont du Gard and being made conscious of the massive elegance with which its three-tiered Roman arches perform the functions of both viaduct and aqueduct.

These exotic examples make the point, but everyday objects, such as cutlery and tableware, when well-made and satisfying to handle, can compel a sense of pleasurable surprise at their quality.

It is in the form of commonly used products that design has its most intimate influence upon the quality of our lives in our homes and workplaces and in travelling between them. Great public works of design and technology are important for national morale and prestige, but we may find our most sustained pleasure and sense of identity in kitchens, on tables, in our furniture, clothing and personal possessions.

In producing this kind of response in the user, design demonstrates its capacity to elevate necessary, commonplace activity to a more satisfying level. It may jolt the user into a new awareness of what is possible and thereby raise expectations and standards.

The second kind of aesthetic experience has less to do with a sense of surprise and more with a sense of reassurance to the user's existing ideas and values. Here the well-designed object seems to reassert an existing sense of rightness by providing another object which meets and confirms expectations. It might provoke a comment like 'Now that's what I call a staircase!', 'I simply must walk over that bridge!' or 'How well that tool works'. What matters here is that the well-designed object enables the user to reaffirm existing aesthetic values and expectations, contributing perhaps to what J Z Young calls, 'the most central of all biological functions – of insisting that life is worth living'.

Both kinds of aesthetic experience are of value because they heighten one's enjoyment of what can be achieved and what to try to emulate. Most of us experience the exhilaration of aesthetic responses and find that they colour what we expect of good design.

TRUTH TO MATERIAL

Two criteria, above all, have tended to dominate the evaluation and appreciation of design. The first of these is firmly rooted in respect for the characteristic quality of materials. Throughout the long and noble tradition of craft-led design, appropriate use of the qualities of a material was almost inevitable since craft enshrined knowledge of a material and suitable ways of working in it. The forms of designed objects came about by gradual change. Even where the craft tradition of one country was influenced by another, designs were modified and translated as they were absorbed into the pattern books and working practice of workshops. Examples of this process may be found in the Dutch influences upon building in East Anglia and oriental influences upon Staffordshire pottery.

As a criterion of good design, 'truth to material' held that a well-designed object should be conceived and formed so as to reflect the more satisfying qualities of the material used. Examples of this are most readily found in hand-crafted objects. However, once recognized as desirable, the criterion could be applied to mechanical production.

To meet this expectation, an object in wood, however made, should come to hand and eye in such a way as to provide a good experience of wood, confirming what is known and expected. It is likely to be polished or oiled rather than varnished and to echo in its shapes, the cutting, hollowing or rounding of the tool or hand which produced it. Its form will be simultaneously functional and uniquely characteristic of wood.

The finial ball on a baluster should invite the palm of the hand which grasps it and the banister rail should invite the clasp of the fingers but allow their unrestricted movement up and down. An object in iron, wrought under the hammer, whether hand or mechanical, should also function well, declaring its strength but also the malleability which it had during manufacture.

This kind of proposition for design is based on what was thought to be a proper respect for the ways of nature. This was seen, in earlier times, to include a decorative and ornamental enrichment provided by nature as part of the 'Grand Design'. The aesthetic satisfaction for the user of an object designed to this criterion was mainly of the second kind above; that is to say reassuring, rather than challenging, to existing values and expectations. It is only since Darwin, and mainly during this century, that we have become accustomed to the idea that what nature has designed is above all functional. The rich variety of forms, which we encounter in flora and fauna throughout the world are now known to result from their need to survive. In the absence of such an insight there could be little to explain the more exotic elaborations of nature other than to think of them as decorative. Certainly much human decoration has imitated patterns of nature even when apparently based upon abstractions often mathematical or rhythmical in origin.

FITNESS FOR PURPOSE

Early in this century, the rise of machine production and the pioneering influence of Constructivism in art, have served to make 'fitness for purpose' a popular and dominant criterion for design. It caused the streamlining of dynamic objects such as aircraft and motor cars and influenced other objects which, although static, were associated with symbols of the then modern age of speed. The washing machine and refrigerator made use of pressed-steel production methods, similar to those of the motor industry, and assumed their rounded forms by common inheritance and their air-age symbolism by association.

The introduction of the Chrysler 'Airflow' (the syllables were run together to form a smooth continuum) motor car in 1934, is remembered as a shock to one's perceptions. It was too innovatory to sell well then, though it was closely echoed in Ferdinand Porsche's design for the Volkswagen Beetle, which, designed in 1936, did not have to meet the test of public opinion until after the war.

The other great expression of an updated functionalism has been manifest particularly in architecture and interior design, in the brutal honesty of construction and exposed service systems. Exposition of the function of a building or environment helps to meet a human need to 'read' and relate to the environment. This public need is not met by vast sweeps of anonymous brick, or concrete cliffs, no matter how functional the buildings may be for their owners, nor how satisfying on the drawing board.

The functional imperative tends to lead to uniformity; where objects have a common function, design solutions tend to gravitate towards a standard. This is seen in the familiar high street, where the function of selling the same range of consumer goods to similar sections of the population, results in each high street resembling every other one in every similar town. This is the case at street level where the sales function dominates. There are often interesting comparisons to be made with higher floor levels where different criteria may be in evidence. Each retailer regularly introduces changes, to layout and shopfront, on the basis of marketing principles, but as these broadly apply to most locations, the more they change the more they stay the same. Perhaps this tendency toward uniformity, assisted as it is by internationally adopted technology, brings us to a consideration of new factors affecting design.

Reactions to uniformity

One way of countering uniformity is to design short-lived goods, where fashion and style can prevail as a source of renewal. Fashion can afford to be frivolously experimental, while being serious of purpose. This benefits the consumer by offering wide opportunities for choice which, when exercised, can lead to the birth of 'classics' which survive because they have the capacity to occupy, uniquely, a permanent niche within our culture. At this end of the design spectrum there is a discernible link with fine art; many artists explore new perceptions and materials for their aesthetic potential. They work within a respected tradition, in which artists bring their discipline and skill with materials to the exploration of perceptions and values. In order to succeed, artists must convince their peers, or a subsequent generation, of the value and relevance of their findings. Although it is not its sole function to serve design, art makes continuous additions to the real and imaginary perceptions which design requires.

It is not only through fashion that art serves design. Design and innovation in heavy industry would not continue for long without the regenerative creative researches of young artists.

NEW FACTORS

The esoteric experiments of Picasso and Braque which gave rise to Cubism originated from their interest in how common objects were perceived and represented. Their images of interrelated common objects broke up the surface of their paintings into forms which were more geometric than descriptive. These forms could later be associated readily with machines and the exciting expansion of the means of production. These were stylistically incorporated into engineering and domestic design through the 1920s, since when they have enriched the vocabulary of design. Bauhaus artists Moholy-Nagy and Josef Albers pursued an understanding of the effects of minor changes of colour and form and to do so used geometric forms and near-industrial production methods. Their influence survives in the kinds of research methods used by artists and designers who seek a rational approach to understanding human responses to form and colour. Such influences have been much more than a jolt to fashion; they have profoundly influenced philosophical, scientific and psychological perceptions of what is possible. It is within the climate created by such perceptions that industry itself can develop successful design and imaginative production.

NEW POSSIBILITIES

Fashion, in dress and textiles, rapidly consumes new forms, textures and colour combinations which are often affiliated with experimental artists. Perhaps the most notable example of the influence of sculptors occurred when, during the 1950s and 1960s, they were amongst the first to exploit the potential of new materials including acrylics, polyesters and epoxy resins. These materials gave rise not only to quite new forms but also to a completely new use of colour in three-dimensional design which had been largely ignored by earlier sculptors. The combined force of their new perceptions and the availability of new materials dramatically changed the appearance of interiors, equipment and furniture design. No doubt the practice of training artists and designers together has helped one to influence the other but, perhaps more important than such direct personal influence, is the fact that the researches of artists who collectively break new ground, provide new terms of aesthetic reference upon which designers and entrepreneurs may draw.

Pierre Teihard de Chardin wrote:

'the history of the living world can be summarised as the elaboration of ever more perfect eyes within a cosmos in which there is always more to be seen .... To try to see more and better is not a matter of whim or curiosity or self-indulgence. TO SEE OR TO PERISH is the very condition laid upon everything that makes up the universe, by reason of the mysterious gift of existence.' *(1959, p31)*

New factors affecting design

International markets and finance

However, it would be equally hazardous for design to look solely to the stylish innovations of fashion, or even to art for guidance. New factors now demand that design draws upon a much wider range of expertise and information. High amongst the new factors affecting design is the increasingly international nature of markets, manufacture and finance. We are frequently reminded, for example, of the need to design for success in the American market for British films: otherwise the necessary international finance cannot be raised to cover the increasing costs of production. In the meantime, American companies, producing international 'blockbusters' are happy to employ British directors, studios and established actors in the design and production of their films. Other industries are affected differently but none is immune to the influence of international markets and finance. The fortunes of the film industry, however, are rather more visible to the lay observer than are the fortunes of many other enterprises.

Successful designers in most fields are quite likely to find employment abroad or for overseas companies and markets. At one time the main reason for this state of affairs might have been that some overseas managers were more design conscious than their British counterparts. The main reason may now be that fewer companies operate wholly within the home market. It is increasingly difficult to cover production costs without a large overseas market for the service or product provided.

More exacting standards

Fitness for purpose is still a requirement of good design but it can no longer be achieved by rule-of-thumb methods. Purposes have become harder to define. There was a time when, working within an assured home market, a leading firm could design a single-purpose tool, by 'passing round the office' two or three mock-up versions and taking a vote, amongst the staff, on the one most likely to sell.

Designing now for a world market is a different matter. Traditional uses, even of hand tools, vary from region to region. New designs do not, therefore, commend themselves equally to all sectors. The way new designs are presented must reflect the regional variations gleaned by market researchers. However, designers themselves must undertake their

On the surface, there can hardly be a more straightforward set of problems in mechanics than those in the design of a motorcycle engine. Japanese engines, like the one shown here, overtook in performance those made in England; but what contributed more to their market dominance was the aesthetic unity of engine design within a new image of the motorcycle. The development of British motorcycles, in which all the parts were improved by a series of modifications, had led to a piecemeal appearance. Successful design has to acknowledge functions which include but go beyond, the technical.

own fundamental research. This means employing methods from both science and art to predict human needs and prepare to meet them.

HUMAN FACTORS

There is ample evidence that otherwise good designs can be rejected by potential users because they require too great an adjustment to their lifestyle. These can vary from social habits to long-established personal preferences. In architecture, people gravitate toward social norms and may reject a covered way between buildings in preference for a muddy track, or may prefer the locker-room to the 'purpose-built' club room as the place to gather for conversation. In the design of medical equipment, the life adjustments required to make use of a valuable prosthetic medical aid may lead to its rejection by the patient.

These examples merely show that a simple statement of purpose, such as 'to replace a limb' or 'to provide hearing for the deaf' can prove to be an inadequate one. Good design questions its own assumptions about purposes. This it must do, not so much by a leisurely discussion, useful though that may be, but by a skilled investigation of the perceptions of the potential user. It is not only the designer's or the client's purpose which has to be taken into account but also the potential user's way of thinking about their purpose. Survival has always demanded an accurate response to the niche one intends to occupy. In design 'to see or to perish' may include the ability to see through the eye of the user.

NEW AND MORE COMPLEX PURPOSES

Another major change for the designer is the sheer complexity of the products being manufactured. In the past, designing a single-purpose tool required knowledge of how it was to be used, by what kind of person and in what range of situations. Today, however, a 'simple' power drill may offer a number of modes of use: fixed, gear-switched or continuous trigger-variable speeds, automatic torque feed-back and response, reverse or hammer drive, pedestal, single or two-handed operation, cordless, power unit or mains supply. All of these alternatives have to be designed for, which means that their uses in real situations have to be envisaged well in advance. Hunches may be a useful starting point but, in view of the scale of investment, accurate research is essential. In making the functions and controls of a multi-purpose tool evident to the user, for example, how is it possible to meet the very different needs, of both the new and the highly skilled user, in the same product? A successful answer to this question may clear the field of competition.

With the rise of electronic controls and information storage, few tools are being designed with only one function. In the design of a multi-functional tool from a camera to a fax machine, or from an office workstation to an aircraft instrument panel, the demands made upon the knowledge and skills of the designer are increased. As discussion surrounding an aircrash or rail disaster always reminds us, design decisions are often of life and death importance. Many devices designed to save energy and labour, demand in return a high level of attention from the user. It is much harder to do oneself an injury using hand-powered garden shears or lawnmower than with powered trimmers and mowers. Complexity and safety are not the only factors which demanded sophisticated attention. Artificial intelligence, increasingly built into software design, inevitably assumes a particular level of intelligence and attention on the part of the user. That assumption needs to be well-judged and the user made aware of it.

New design assumptions
Apart from the problem of how design communicates these assumptions to the user, there is the more fundamental question about the bases upon which design assumptions are made. Human factors cannot be taken, responsibly, into account unless they are properly investigated. Ergonomics, once characterized by data on human physical norms, has considerably widened its scope. It now attempts to meet the designer's growing appetite for better information about the user. Assumptions based on physical or psychological norms can be, in practice, very wide of the mark, and can therefore be dangerous and costly. In view of the risks to which a user may be exposed, and the high cost of misdirected investment, fuller information is needed. A proper study of potential users, their tasks and environments, conducted before design assumptions are made, is likely to be not only cost effective but also stimulating for a design team.

Recent research undertaken by design consultant Fitch RS in the USA revealed that ironing clothes is no longer a part of the laundry process but a task associated with getting dressed. This new and tested perception provoked the design of a one-hand, fold-away, ironing board which can be hooked to any clothes rail, within a wardrobe or elsewhere, and used and stored in that position. It has been hailed as a perfect answer for the compact urban lifestyle *(International Design July/August 1990, Annual Design Review and Platt, 1990)*.

Information technology and video imagery together, make possible simulations of reality which are already very important in training astronauts and pilots. They have, in modified form, made an impact on leisure and education and can be expected to move into greater prominence in the future. We are increasingly required to communicate with machines. Modems, fax machines, computer screens, videos, voice simulators, recorders, paging devices and robots can hold our messages and convey them to others on our behalf, or carry out our instructions, on time, weeks

Materials in design and technology

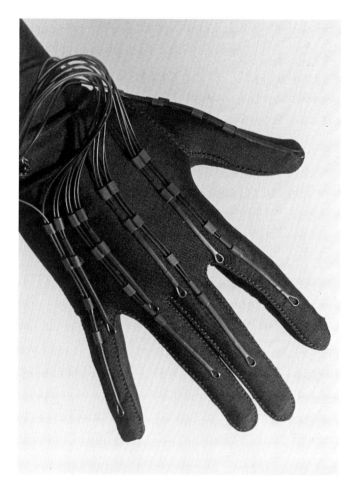

'Virtual reality' is the term used to describe the high resolution, illusory, three-dimensional computer images which can be 'entered' by placing a suitable input device between the observer and the illusion. The glove depicted here allows its wearer to 'pick up' and move objects in the illusory space generated upon a computer screen. The quality of the 'virtual' worlds we create will depend upon the quality of our perceptions.

later; mostly to our advantage. However, such devices can reach into periods normally reserved for travel or for rest and may intensify the stress which they are designed to reduce. Recent battle reports have shown, very dramatically, how remote controlled and 'smart' devices have the capacity to remove us from the material and human consequences of our actions.

All this means that responsible design must take into account the nature of the interfaces between humans and machines. Not only do we affect and control machines; they also influence our lifestyles and exercise control over us. We may now be facing something of a cultural watershed since not only has the need for face-to-face contact been reduced by information technology but so also has the opportunity for the kind of hands-on experience of materials which have formed our standards and structured our language. Recently alleged evidence of decline in reading ability, if confirmed, may or may not be attributable to teaching methods. Old yardsticks make poor divining rods. What we may be seeing here is part of a shift away from face-to-face conversation and reading, towards coping with simulated environments and visually imparted information. Certainly most of us are impressed by the rapidity with which children learn to interpret images on television and computer screens. Whether they learn to do so critically is a matter for parents and teachers. Familiarity with the problems encountered and processes used in designing for these media as part of their education will help to ensure that they do.

New realities

The 'virtual realities' of computer imagery already permit users to experience the illusion of entering and moving about in computer generated environments and to handle 'objects' and 'persons' within them. As technological development makes these illusions ever more convincing, competition for attention will occur between the synthetic and real worlds, and moral questions will undoubtedly arise. The ability to recognize and to understand differences between the real and the depicted world will increase in importance. Technology is sometimes said to be value free. However, applications of technology always involve value judgements, concerning outcomes and effects. Design is never value free and moral questions are never far from practical ones. Again, developing the ability to recognize and understand differences should precede the need to make judgements. Developing the ability to see and to discriminate between alternatives will remain an important objective for education in design and technology.

New materials

Finally, design is faced with the fact that materials themselves can now be designed. New materials can be synthesized with a versatility unknown in natural ones. An effect of this is to remove, to an extent as yet unknown, the limitations of materials to which we have grown accustomed. Electronics and black-box technology largely remove from view the reassuring logic of moving parts. Many of our customary constraints, in function and structure, are disappearing. A rising question for designers and their clients is: 'As we overcome these limitations, what criteria and principles will inform our design?'

Whatever its outcomes, we shall require technology to meet properly assessed needs, to function well, to nourish our spirit and preserve the welfare of the planet. Design cannot meet those demands without the participation of all disciplines and sources of knowledge. Evidence that designers are aware of that need can be seen in the increasing use of design teams consisting of representatives of many fields of study and expertise.

In the academic world at large, however, there is not yet sufficient awareness, within the separate disciplines, of the demands which design must make upon them. It may be that this development will first occur within schools where the need for cooperation between specialists is already obvious and where, not least in primary schools, joint enterprises have already shown their value, in the quality of teaching and learning which they provide.

8 DESIGN EDUCATION

BACKGROUND

Funds for design

'When the British government decided to devote funds to the School of Design, in 1837, it was only a couple of years after they had given the first grant to any English educational establishment. And, as a network of "branch schools" was established steadily afterwards, in major towns and cities all over England, each of them linked to the School of Design in the centre, the national system of art and design education in fact pre-dated the national system in any other subject.'
(Frayling 1987, p12)

Frayling goes on to point out that one effect of this early interest by government in design education was to set it apart from education in other areas. That effect lasted well into the present century and it is only now that design is beginning to receive high-profile attention at all levels of mainstream education.

Then, as now, government interest in design arose because of the need to improve the competitive performance of industry. Then, as now, there was much discussion about what constituted a 'proper' education in design. It was quickly linked with art and craft and flourished in some respects amongst those who sought vocational education and employment in this specialist area.

However, institutional isolation not only denied the subject to students in general education but also limited the access of its own students to other subjects. It is not surprising therefore, that for many, the use of the term 'design education' needs some explanation.

WHAT IS DESIGN EDUCATION?

The term, despite its academic resonance, does not yet refer to a fixed entity. It has been in professional use for some years but what it has referred to has changed with time and circumstance. It may be associated with a wide range of teaching styles, from problem-solving to technical instruction, or to an expedient grouping of subjects, or to the content of a course. The attachment of the title does not in itself confer authenticity upon what it describes.

Nevertheless, the term has a legitimate use which is to describe a movement generated and sustained by teachers who, coming from different backgrounds and training, have sought to increase recognition of the educational importance of design. Their commitment has focused upon design awareness as an objective for education and upon design-related activity as a contributor to general education. In the face of discrepancies in the use of the term it seems reasonable to expect that what it describes should be educational and should either involve the learner in some aspect of designing or lead to a greater critical awareness of design.

Because the authentic practice of design involves active use of the best knowledge available, design education in schools should draw upon whatever relevant experience children can bring to it. Some of this knowledge is acquired out of school but much of it must come, as required, from whatever part of the curriculum can provide it. Its characteristic appetite for omnifarious knowledge makes design difficult to place in the traditional, inherited school programme of discrete subjects. Curriculum planning or time-tabling, which is often based on attempts to arbitrate between the different requirements of separate subjects, tends to perpetuate the limitations of that inheritance.

The day-to-day matching of children's activity and attention to what is actually done continues to depend on the ingenuity of teachers and their considerable skills. For whatever reasons, cross-curricular teaching is more prevalent in primary schools than in secondary schools and more facilities exist there for design education to draw upon relevant aspects of children's learning. Encouraged by the National Curriculum, secondary schools will no doubt find ways to provide broadly-based teaching in this area but design is not yet seen as essential to the teaching of all subjects.

Currently, design education is changing and is different from one local authority to another and from school to school. This is by no means an unhealthy condition for a developing aspect of education, and enterprising authorities and schools make strong, educational responses to the requirements of National Curriculum Design and Technology.

Overt developments in design education have largely occurred in secondary schools because of the subjects which have so far contributed most to it.

These will be outlined later but it is first necessary to take note of independent developments in primary education which, although not identified as design, nevertheless prepared the way for it by puposefully involving children in activity with materials.

DEVELOPMENTS IN PRIMARY EDUCATION

During the period between and immediately following the European wars (1914-18 and 1939-45), a revival of the art and craft movement made a strong contribution to the development of primary education in Britain. Instead of regarding children's work merely as unskilled striving after adult imagery, informed teachers were now able to understand that the characteristics of children's drawings and paintings depended at least as much on how children saw and thought as upon their dexterity. Teachers were, for the first time, able to identify and respond to some aspects of children's work which had previously been undervalued.

These important insights developed from pioneering work by teachers who had been influenced by developments in art. Changes in art, beginning with nineteenth-century Impressionism, broke down the dominance of academic realism and allowed for the development of many alternative ways of describing and depicting reality. What was new and valuable to teachers was the re-emergence of many alternative models of reality produced by adults from within their own culture and from others which authenticated the potential of propositions made by children. The validity of visual and tactile vocabularies used by children, now taken for granted in graphic and product design, had not previously been widely recognized or encouraged in education.

The developments within art itself led to two great contributions by twentieth-century art and design, namely Expressionism and Constructivism. The first emphasizes the experience of the individual and the second emphasizes the elements which we manipulate to establish orderly forms and functions. Both of these complementary traditions have contributed to the vastly increased range of models which we may use to understand the propositions made by children. Their combined influence in the practice of primary teaching has meant recognition, by many teachers, that children learn by making propositions about things and then testing them against their own experience. Although this valuable insight began in respect of children's drawings and paintings it rapidly spread to include their writings and general perceptions.

New responses from primary teachers
As teachers demonstrated their response to their ideas, children worked with impressive energy and often produced work of high quality, with acute perception and emotional power. There are historical reasons why

Design education

the pioneer teachers were significantly influenced by the expressionist half of the tradition and why, so far, it has had the stronger and more confident response in education. It led to that vital generation of creative activity with materials which has nourished so much of the best primary practice since the last war. The best practitioners, amongst teachers, have always understood that expression is founded upon investigation and knowledge. Some of the strongest work in this tradition arose from local studies or field studies of a cross-curricular nature.

> 'The achievements of many junior schools in producing working models of great sophistication, three-dimensional sculpture of remarkable originality, and a highly competent range of craft products in clay, cane, paper, card, and wood is well known and an established part of the curriculum of the first school. And the integrated project activity, for which many of the schools are famous, regularly and successfully involves the use of materials.' *(Eggleston 1976, p4)*

Technology can mean the making of working models from kits which are designed to meet the kit-maker's objectives. They are also often designed to reduce difficulties for children and consequently reduce their opportunities to encounter mechanical principles in real situations. Here, a child has made a vehicle to carry a brick down a slope and is adding extra features.

Although children's creative work is often artistically powerful, there are many other factors in the propositions they make; including geometric, arithmetic, constructive, abstract and technological factors. Not all drawings and models, whether made by children or adults, are regarded as 'art'; many have more to do with engineering, plans for buildings or journeys, designs for apparatus and equipment, often needed to give substance to imaginative ideas and invention. Education has, so far, been much less confident of its response to these elements in children's work.

Misconceptions are highlighted by such inadequate phrases as 'junk modelling' and 'self expression' which have not been helpful in education. Materials cannot properly be described as 'junk' if they have potential, and the term can only be derisory either of the material provided or its resulting work. 'Self-expression' is similarly not a suitable term to encourage work of any quality, or even of individuality, since it fails to acknowledge the importance of ensuring that the 'self' acquires experiences and ideas worthy of expression.

Indiscriminate praise or approval of children's work is of little value in the short term and in the long term is counter-productive. Since no model, drawing or description equates precisely with reality, what matters most during learning is that what the child proposes should be understood. This requires some effort, skill and persistence on the part of the learner but also some familiarity, on the part of the teacher, with the uses made of drawing by artists and designers, including engineers and architects, because children will use drawing for similar purposes.

In case studies of 20 designers, including architects, ceramicists, engineers, silversmiths, fine artists and graphic designers, Steven Garner *(1990, p40)* found drawing used for a variety of purposes: as a means of learning and communicating; of problem finding and solving; of exploring and manipulating ideas. He quotes product designer Dick Powell as believing that:

> 'One is more of a creative person if one can draw ... because you can have this conversation with yourself, you can express your ideas to others and you can organise your thoughts better.'
> *(Garner 1990, p42)*

Dennis Atkinson *(1991, p62)* points to the range of such discourses for which children use drawings and emphasizes the need to understand how drawing functions for the child.

> 'The form of a child's drawing may appear simple, primitive or deficient in some way, to those whose judgement is affected by a particular attitude towards representation – intellectual realism, for example. But in simply considering the form we are likely to miss the crucial factor, that is, how the drawing is functioning for a child in his or her particular drawing discourse. This has implications for assessment procedures in teaching, which neglect the functional significance a drawing has for a child.' *(Ibid, p71)*

Appreciation of ideas
There is no more encouraging experience for learners than to discover that their thinking about what they observe and describe is understood by their teachers. Teachers who can demonstrate that the thoughts which children convey are received, are likely to teach successfully. Whether the means used by the child is mainly verbal or graphic, genuine appreciation of the idea by the teacher provides a good basis for teaching. This is not always possible, but where adults are aware of a range of ways of seeing and of representing reality they are usually more able to understand the nature of the proposition being made by children. The relevance or usefulness of a model cannot be evaluated or discussed with its proposer until what is being proposed has been substantially understood.

The following example may illustrate this in a practical context.

A child in a reception class has used pencil marks and paint to represent a bus. The drawing is on white paper approximately 25 x 20cm and consists mainly of rectangular shapes, only some of which are painted red. It is reasonable to suppose that these correspond with the pattern of windows and body panels observed on a bus. This is confirmed when the teacher notices a face drawn in one of the less densely painted rectangles. This, the child explains, is the bus driver. When the teacher asks if there is anyone else on the bus the child affirms that there are passengers but that they are inside. The point of this explanation becomes clear when the child turns the paper over to reveal the inside of the bus, which is represented by a series of smaller rectangles, each of which is occupied by a drawing of a face.

Being able to understand, and therefore to welcome, this achievement the teacher went on to discuss with the child the question of how it would be possible for others in the class to enjoy the drawing. Since the display of work on a wall was the norm for this class, this particular drawing presented a problem. It was solved to the mutual satisfaction of teacher and child by propping the drawing up in a wire book-frame on top of a bookcase where it could be seen on both sides.

This is an everyday application of teaching skill but one which exemplifies the importance of a teacher's understanding a proposition before responding to it. With older children the same principle applies but the way of reacting to their proposals can vary with the kind of encouragement needed and can include whatever suggestion for development is deemed appropriate in the context of the work.

Since children depend on their powers of perception and description for their approach to learning, their success in using them should be recognized whenever possible. Better perception and clearer expression is always achievable by those who retain confidence that they are equipped with the necessary powers. Those who lose that confidence may resort to less appropriate means to overcome inadequacy. Many adults are reluctant to draw simply because, during their early education, no one succeeded in demonstrating to them the validity of their early drawings. This response applies to all kinds of modelling and structuring of ideas in materials.

'The range of work is vast: the history of many of the activities is as old as mankind itself. Yet only recently have we begun to realise the full potential of this area of the school curriculum. Not only have we discovered a wide range of new and previously unused materials; but also we have rediscovered the intellectual as well as the practical learning that can take place in work with materials.' *(Eggleston 1976, p1)*

What Eggleston says about drawing applies equally to the full range of materials which are available as an aid to discourse. Howard Gardner refers to the work of Claire Golomb with children using three-dimensional media as demonstrating striking differences in concepts and levels of integration according to the media used.

> 'Golomb's findings clearly refute the theory that the child's cognitive level imposes itself in all tasks, regardless of symbolic medium. The child's conceptual level must interact with the experience of and the demands of each medium.' *(Gardner 1978, p344)*

Teachers are aware that the work that children do varies in purpose and confidence with the nature of the materials which are available for them to use. Sally Atack *(1980)* describes some of the benefits for handicapped adults and children of activities with a range of materials. In an earlier unpublished study involving 1,100 utterances from 54 pre-school children, she observed an apparent link between certain materials and concepts, indicated in words arising in their discourse.

The accumulated insights of teachers are important and contribute greatly to the ethos of schools and to the quality of learning which children experience. A task for those seeking to develop a sound foundation for design and technology in primary education must include finding ways of valuing and effectively sharing the expertise amongst primary teachers both as it exists now and as it continues to grow.

SECONDARY EDUCATION

Craft, design and technology

Since 1945 there have been many attempts to develop design education in secondary schools, but they were generally linked either to art – as in art and design – or to craft – as in CDT. However, since design cannot function properly where it is exclusively linked in either of these ways, such attempts have served mainly to reveal their limitations.

Craft, design and technology, or CDT, as it became known in secondary schools, was a post-war phenomenon which grew out of dissatisfaction with what were then known as the practical subjects of woodwork and metalwork. These had been developed towards the end of the nineteenth-century with a view to providing 'manual instruction' for boys, with a similar pattern in domestic crafts for girls. In optimistic post-war conditions these subjects were seen to be less appropriate to life in a modern industrial economy. They were seen also by children and parents, as being concerned with the learning of practical, rather than intellectual, skills and consequently were considered to be of relatively low status. A further subject, usually taught in the same department as woodwork and metalwork, was technical drawing. This gained some status by the association of its standards with continuing engineering drawing practice. Teaching of these practical subjects only rarely provided opportunities for children to engage in design. Teaching was most often directed towards pre-determined outcomes which included additional skills. The complex range of skills taught to girls was gradually rationalized within a more theoretical framework and called domestic science.

With the need to meet the demands of comprehensive education and to provide courses appealing to children of a wide range of abilities, and later of both sexes, there were further good reasons for radical change. New courses were devised to include more opportunities for design and for acquaintance with new materials, techniques and processes. They were identified by the new titles of craft, design and technology (CDT) and home economics. These required a move away from the long-established workshop practices within which most teachers of the subjects had been trained. Many were reluctant to depart from the security of their training and some were in any case unconvinced of the need for change.

Other teachers, grasping the opportunity, were keen to provide new teaching to support the children who were often asked to solve much more challenging problems. While retaining the best of the craft traditions, they eagerly undertook to teach electronics, principles of design, graphic communication, food and materials technology or indeed anything else that children needed to know in order to engage seriously with design and technology. This demanded some courage and considerable diplomacy on the part of CDT and HE teachers, since they were soon teaching in areas outside their personal expertise and encroaching upon the territory of specialists in other subjects.

It is partly in recognition of both the achievements and difficulties of teaching these subjects that they are now being seen to lie within the framework now afforded by design and technology.

Art and design
Because of the different history of the way art and design teachers have been trained, most of them have long been aware of an association between art and design. Most would confine their teaching of technology to those techniques, materials, and processes which are directly related to art and design production. The local and regional colleges of art where most such teachers underwent their initial training, had been set up in the 1830s as schools of design, when there was a perceived need to improve the quality of design in manufactured goods, particularly in textiles (see the quotation from Frayling on p78).

As we have seen, any attempt to teach design inevitably demands a wider curriculum, and the design schools were no exception. As they

Perception can stimulate technology which can in turn stimulate new perceptions. Here a student, accustomed to weaving flat surfaces, wanted to describe the form of a leaf which undulated three-dimensionally. She devised a loom which could present a warp at a number of levels. Because the structure depended upon tension in the warp, the woven leaf could not be cut from the loom; so the loom was constructed from a transparent material and left in place as a frame for the work. This kind of interplay is a constant factor in design.

responded to later nineteenth-century developments, they found themselves becoming schools of art and crafts, with a strong emphasis on drawing and the study of fine art as a source of standards. Fine art and crafts dominated their curriculum, until well into the twentieth century, but in 1946 a National Diploma in Design was established. Since then there has been a steady development of both higher diploma and degree courses in many specialized aspects of commercial and industrial design in regional colleges throughout the country. Many of these courses are now within polytechnics.

As a link between general secondary education and specialized art and design degree courses, most undergraduates undergo a preliminary diagnostic course, of at least one year's duration, in their local school or college of art and design.

The existence of this kind of preparatory course may have militated against the widespread development of strong design courses within secondary schools. Sixth-form courses have not been required to adjust very much to the changes in colleges of art and design mainly because colleges have preferred to undertake this preparatory and diagnostic work themselves. Furthermore most teachers of art and design in secondary schools have been recruited in the past from training courses in fine art. Many have felt that their first responsibility was to focus attention upon aesthetic, rather than academic or economic, criteria.

Two further reasons may account for a period of inertia in schools. The first is that teachers of art and design in colleges historically had very little influence over the examination system, dominated as it was by requirements of the universities. Their examination boards had some familiarity with the history of art and architecture, as academic subjects, and with fine art within one or two universities but, until very recently, they had no familiarity with design.

These factors gave very little confidence to teachers in colleges of art and design, in the value of A level examinations for predicting success in college courses. Given evidence of a reasonable balance of subjects in 16+ examinations, teachers in further and higher education would place much greater predictive value in a personal examination of the applicant's folio of practical work. There was then, at least, some likelihood that selectors could identify students able to benefit from their course.

The second reason lies in the nature of the course or courses that colleges can offer. The most immediate evidence of departure from normal school traditions is the extension of time devoted to sustained practical work. Much of this time is devoted to the exploration of materials and processes in very challenging conditions which have not, traditionally, been available to sixth formers in school. Aptitude for highly specialized degree courses could not be identified on the restricted range of practical experience provided in many school sixth forms.

The higher education courses in design, to which local colleges provide a transition, include the following specialized opportunities: advertising and promotional graphics; book illustration; technical and scientific illustration; print; typography; photographic, television, video, and computer graphics; three-dimensional design; product design; interior design; furnishing; theatre; glass; jewellery; conservation; packaging; fashion; textiles; footwear; ceramics; exhibition and museum design; film set design; design for crafts, with new courses replacing the old as the economic environment requires.

Separate development
Head teachers are often distinguished academics. Traditionally, schools favour academically-led careers for able pupils and have not always enabled them to see opportunities for personal fulfilment in design, technology or other aspects of economic life. While many able students go on to careers in industrial management, for example, the educational route which they have been required to follow in pursuit of an academic degree has often steered them away from all direct practical work with materials or with technical processes.

Additionally, students who demonstrate academic ability are readily identified as able, while those who demonstrate artistic, creative or practical ability are often thought of as possessing something called 'talent'. Failure to recognize 'talent' as an artistic or practical ability, which needs to be challenged and educated in the same way as academic ability, results in its loss. This is not a charge which can be laid entirely at the door of the school, for the myth that 'talent' is inborn and somehow not amenable to education is widespread. It is important to attend to these abilities and discover what kinds of persistence and discipline are necessary for their development. It is to the credit of the working party for National Curriculum Technology that it has identifed and described a 'design and technology capability' towards which all children are entitled to progress.

Attempts at reconciliation
In the 1950s a major initiative in the colleges came to be known as the 'Basic Design' movement. It had its origins in Constructivism which sought to identify elements such as the point, line, colour and space which make up every work of art, whatever its particular form. It attempted to provide a rational basis for teaching a grammar of visual form which proved useful to advanced students and, perhaps surprisingly, to fine artists. Through them it had some impact in schools, enabling more art

teachers to draw upon their students' analytical abilities and to help them understand the formal elements of visual imagery. Teaching grammar is as important in teaching art as in teaching the mother tongue. However, in both cases a degree of fluency and familiarity with the uses of language as a whole, is required. In the limited time available for art teaching in schools, basic-design teaching paid too little regard to the need of students to learn by using the whole language independently, in ways which were meaningful to them. Of inestimable value in teaching the few, it nevertheless tended to limit – rather than facilitate the development of – the many, whose thought processes appear to run on different lines.

The overriding point to be made here is that neither art and design on the one hand, nor craft, design and technology and home economics on the other, were able to provide an adequate home for teaching design, because both approaches were weighed down by the limitations of their parent subjects and traditions.

During the 1960s and later, a number of local authorities took the opportunity, when reorganizing schools, to purpose-build design departments, or to group practical activity-based subjects within a design faculty. At their best, these facilitated cooperation between teachers and allowed some to teach as a team, offering relevant expertise as and where required. Much depended upon leadership and the recruitment of experienced and committed staff. There was then the task of setting up sound departmental policies and common teaching principles; these took time to establish and did not always survive changes of leader or team membership. There has probably been no recent educational development which has depended more on the tactful support of local authority advisers. Less successful departments sometimes fell prey to time-table expediency and 'circus' arrangements which trivialized work and destroyed continuity.

Partial success
Neither buildings nor administration guarantee success; there have always been inspired and brilliant teachers who could teach aspects of design and technology effectively by simply breaking out of the limitations of their subject. Sometimes this was achieved by exceptional teamwork involving teachers who could, together, draw upon wider experience. Occasionally, charismatic individuals, often with unusual backgrounds, were able to provide good experience of design for their students aided only by the force of their personal convictions and knowledge. Some of the best teaching and most rewarding outcomes occurred where students were able to combine the benefits of creative art teaching with the benefits of technical knowledge derived from effective CDT teaching.

However, none of these sources of better practice could provide a general model for teachers who were unable to break out of the limitations of their subject especially where these were reinforced by the conventional expectations of examinations and timetables. The absence of a rational framework for the teaching of design meant that the general level of practice often failed to fire the imagination of the students or to make sufficient demands upon their capacity for designing. Outcomes were, then, of poor educational quality, not related adequately to the actual abilities of students nor to the realistic requirements of professional practice.

In these circumstances neither teachers of art and design nor teachers of CDT and HE were satisfied. Some were tempted to avoid the pressure to teach design. Members of both groups felt obliged to defend their parent subject from what they saw as dilution by the different, and often undesirable, values of another subject.

Sustained professional distrust in teaching is both arduous and painful and is unlikely to be resolved, to the advantage of students until a sound rational framework is achieved. The National Curriculum for the teaching of design technology is a step in the right direction in that it has made clear the expectation that it will require input from teachers of a number of subjects if it is to succeed.

Children enjoy taking things apart to see how they are constructed, and 'reverse' engineering and designing are respectable industrial practices in which a competitor's product is dismantled to see how it is assembled. The structure of this staircase demonstrates how the turn of the treads saves floor space.

New possibilities

Success will remain only partial until we acknowledge that design must draw upon all available knowledge and disciplines. The working party, which drew up the National Curriculum for Design and Technology, reported that this point had been made to them by representatives of the design profession. The working party had felt that although a cross-curricular approach would not daunt primary teachers it would be too demanding for the specialist climate of the secondary school. The tactical need to establish technology as a discrete subject, in secondary education, will militate against the long-term interests of design if it delays recognition that fostering capability in design is a responsibility to be shared by teachers of all subjects. Similarly, capability in information technology, included within the attainment targets for technology, will need the support of all teachers.

Although the working party for technology alone was unable to tackle the traditional inertia of secondary education, the final statutory requirements for technology make clear that it must include substantial experience of design for students. Non-statutory guidance produced by the National Curriculum Council and the Curriculum Council for Wales, at the request of the Secretary of State, recommends that 'teachers of art and design, business education, craft design and technology, home economics, and information technology/computer studies have vital contributions to make.' *(NCC 1990, p310)*

This is a reasoned compromise but it falls short of asking teachers of all subjects to include design-related activities in their teaching. Hopefully, this will be achieved, although for many secondary school teachers, a positive response would require a profound change of teaching style.

FERTILE GROUND IN PRIMARY EDUCATION

No such problems exist for primary school teachers, who are mainly accustomed to a teaching style which helps children apply new knowledge in work leading to practical outcomes.

It is intended that this brief outline of some factors in the development of design education should make clear to primary teachers that all the understanding of the subject does not lie with the specialists in the subject. As we have attempted to point out, specialists are generally enthusiasts for aspects of design education which are closest to their interests and training. Primary teachers have new ground to break and they should not allow the historical difficulties encountered in secondary education to be visited upon their schools.

Advice from any quarter, including this book, should be viewed critically in the light of what teachers know of the needs of primary pupils and of particular schools.

Much of the best of primary practice already ensures that children have some contact with design-related activity and with technology. This was observed and acknowledged in the Design Council's Primary Education Working Party report:

> 'Because design involves making choices about our environment, teaching which develops children's awareness of the character and organisation of the natural and made world, and encourages them to respond to it, is design related. So is teaching and learning which develops children's perception and understanding of forms and structures, and of how their arrangement is related both to physical

laws and to their purpose and function. So is activity which gives children the opportunity to manipulate selected materials in an expressive and imaginative way, and to experiment with such things as colour, texture, and form.' *(1987, para 3.6)*

The report goes on to show in some detail the scope for design related activity across all the the subject areas of the primary curriculum.

With more knowledge about the practice of designers and technologists, primary teachers will be able to apply their existing insights to developing a broadly-based design education. It is quite important that where schools adopt any of the practices of professional designers they understand how these practices function for designers and for their students. Mere imitation of some of the methods of professional designers will be both misleading and counter-productive.

One of the ways for teachers to acquire this knowledge is to explain their needs to the design departments in their regional colleges of art and design, polytechnics and colleges of further and higher education. Here they can discover how designers are trained and how they work. These departments throughout the country constitute a major national resource, which has not, so far, had many calls made upon it from primary education. There is evidence that where well-considered requests have been made to individual colleges, responses have been both encouraging and enlightening.

Schools need to be imaginative in exploring local, regional and national agencies who can facilitate contact with professional designers and technologists. By informing good teaching with reliable knowledge of the work of first-class designers, the necessary enterprise of establishing design in mainstream primary and secondary education can become a collaborative and rewarding venture for everyone involved.

9 MATERIALS AND MEANING

THAT materials can convey meaning is, on the face of it, something of a mystery. The spiritual uplift which can be experienced in a cathedral would not be possible without the stones and timbers from which it is built, nor would the humanity seen in a Rembrandt portrait be achieved without his choice of finely ground pigments and oil.

We depend upon very specific arrangements of particular materials to give us access to such feelings and ideas. It is not the material alone which produces meaning but the persistent actions upon it of many people who have worked to discover and realize its potential. Without the materials there could be neither cathedrals nor paintings, and without generations of human endeavour with materials, we should not have achieved any framework for concepts like 'spiritual uplift' or 'humanity'. It is only our exposure to the many models or embodiments of such feelings, in our language and all other media, that enables us to have shared access to such 'abstractions'.

THE INSTINCT TO LEARN

The instinctive responses which children make to the material world do not vary much from child to child. One such instinct involves using certain material objects as playthings and attributing special significance to them. All children do this but the way they do it varies. As they gain experience the range of objects used may increase and so too the things they represent. The meanings, which objects and materials embody, appear gradually to extend and are incorporated into play with other children.

Instinctive responses are an aid to basic survival. However, in order to advance their culture, higher species need to make more complex forms of adaptation than inherited instinct can provide and these responses have to be learnt by each individual within a culture. In order to be useful such learning must include socialization. This means developing a capacity to make shared responses, and equally a capacity to make individual responses which can contribute to cultural advance. It is by the interplay of these two aspects of learning that the human species has advanced beyond the limitations of instinct. By using language and other media, individuals contribute their learning to the stability and security of the culture and also to its capacity for change.

Writing in 1896, Karl Groos points out that intellect alone can accomplish more than instinct and that, therefore, individual experience becomes more and more prominent. He notes that it is in the higher orders of animals that play makes an appearance in order to overcome the 'swaddling of inherited impulse':

> 'Now we see that youth probably exists for the sake of play. Animals cannot be said to play because they are young and frolicsome, but rather they have a period of youth in order to play; for only by so doing can they supplement the insufficient hereditary endowment with individual experience, in view of the coming tasks of life .... The animals do not play because they are young but have their youth because they must play.' *(Bruner, Jolly and Sylva 1975, p66)*

This view of the evolutionary importance of play suggests that the extended period of preparation for adulthood in humans has, by allowing for the necessary individuation of responses, made cultural and technological progress possible. Although we have, as Richard Gregory pointed out, escaped our biological origins we have to recognize that it was those origins which enabled us to do so.

Jerome Bruner wrote:

> 'I would only urge that ... we keep our perspective broad and remember that the human race has a biological past from which we can read lessons for the culture of the present. We cannot adapt to everything, and in designing a way to the future we would do well to examine again what we are and what our limits are. Such a course does not mean opposition to change but, rather, using man's natural modes of adapting to render change both as intelligent and as stable as possible.' *(Ibid, p60)*

It may be significant that at the time when the main traditions of our education system were set up in the mid-nineteenth century no one was aware that the human race was more than a few thousand years old. It is only now as we contemplate education for technological change that we are made to see the processes of design as rooted in our evolutionary past.

In their play children learn to take possession of their environment and to extend their knowledge of it. They experience the twin necessities of security, rehearsing what they know, and of adventure, exploring what they do not yet know. New materials and new circumstances are a threat which must be adventurously overcome in order for them to become part of the secure realm.

Children use materials in play for a number of identifiable purposes of their own, each of which provides opportunities for learning.

PLEASE FROM EXPLORING MATERIALS

The first of these purposes is directly for the pleasure which children can derive from handling the material. Materials are used to alert and to delight the senses. As food is eaten with pleasure so it is also examined. Crayon marks are made, lines are drawn, clay is kneaded, sticks are balanced, stones thrown, towers built and demolished for sheer pleasure. Such pleasure is also very often social, so that both the experience and the pleasure in it are shared; it then becomes intellectual, giving rise to speculations, questions and predictions.

Both adults and children like to trace lines upon surfaces as in cycling or skiing, moving through air or water as in swinging, swimming, or diving. A similar type of pleasure is associated with live material: many people particularly enjoy gardening, rearing pets, riding, or birdwatching. An indication of the importance of such 'simple' pleasures is the extent to which we plan for their inclusion in our lives and invest in ways of providing them. Discrimination is required and aesthetic and ethical judgements quickly follow. Parents and teachers are involved and can help children to recognize and talk about experiences which call for adjustment. Circumstances change and when children repeat experiences in the expectation of repeating pleasure they are sometimes disappointed. Children can be helped not only to react emotionally but to look for differences and reasons for the unexpected. The transition from disappointment to curiosity may require adult intervention; this should not attempt to invalidate the emotion but to divert the energy which it generates into further inquiry. Disappointment can often be the result of inappropriate, or uninformed, expectation and is therefore an opportunity for learning.

The younger, or less experienced child, may have very few expectations and is often content merely to discover what a material is like with no further expectations of it. An older or more experienced child may need encouragement to take pleasure in discovery, since this often leads to realizing what can be achieved with a material. Half-formed images or possibilities can be talked about and completed or passed over in favour of others. There is no moral obligation that all activity with material should produce objects worth keeping. Many outcomes are retained to keep the ideas alive as they progress toward fuller realization. Engineers and designers make the practice of having their unfinished projects and ideas on view, around them, to keep them actively in mind and to discuss them with others.

A practice of designers from which children can learn, is the generation of many more possibilities than can ever be pursued. An essential ability for children to develop is to generate ideas without the anxiety that all must be worth following through. This enables them to develop a further skill of selecting those most likely to be worth pursuing.

Playfulness with ideas is a necessary aspect of design and creative ability. Designers tend to file ideas, that is to say to make and store notes for future reference. A too early attachment to first ideas can stop active design in its tracks. Too often as parents and teachers we leap to interpret a by-product of exploration, where we might learn more by watching how the child stores away valuable experience or applies it immediately.

SUPPORTING IMAGINATIVE PLAY

A second purpose for which children use materials is to support their play. Objects which are ready to hand can be used for this purpose. They do not have to be replicas of the real objects for which they stand, but they have to be able to function within the implied rules of a game. It is natural for children to imitate the roles which they see enacted by adults – to play 'at' something. Materials, including bases and locations, are needed so that there can be an identifiable 'hospital', 'castle' or 'office', and to act as props for the play or costumes for the characters.

It is important that the ideas, simulations and pretences are owned and understood by the players. It is a characteristic of children's play that there is much discussion of the 'rules' which are needed to make the game work. Whereas reality is not required, since all participants will suspend disbelief for the duration of the game, certain rules of logic may emerge as essential for its continuance. Leadership, rule-making and rule-following are practised in such play: propositions are made and tested within the protection that play is not 'real'. However, the more their pretence calls for the use of materials the more children encounter problems. The fact that these encounters with real materials occur within the safe constraints of pretence is precisely what gives them their educative cultural value.

Even within such protection the problems raised by materials become real; in the sense that if a structure, for example a tent, is required to function as a house or a discarded packing case as an aircraft, these 'props' have to be capable of carrying the imaginative properties attributed to them. If, for example, a large cardboard box is to function as a cockpit for a helicopter it may be essential that the pilot and passenger can sit side by

Materials in design and technology

These life-sized figures gave a physical presence to two imaginary characters invented by children with the help of a visiting writer. Together they developed a story about children evacuated to their village during the last war. It involved research and local enquiry. They made a collection of wartime memorabilia and reconstructed aspects of the wartime environment.

side and discuss the imagined prospect before them as they pretend to fly. If, however, the same box has to function as a taxi, a driver-fore, passenger-aft, arrangement may be wholly acceptable. In such play the players set the criteria by which they will judge what they make. Where an adequate range of materials and some confidence and skill are available, these criteria can usually be met. Indeed in some play the making of props becomes a major preoccupation and when successful can feed into the play elaborations and refinements not initially envisaged. An equally positive outcome can occur when a game is changed because the first idea cannot be realized in available materials, but others can.

The structural properties of wood or clay, the technology of paint or the behaviour of levers are consistent and present the same face to adult or child. When dealing in their play with such materials the mind of the child has to cope with real properties. Children damming a river in play encounter the same destructive power of water as did their ancestors, and which their own culture will have to deal with repeatedly in the future, and they can design strategies for dealing with it, according to their understanding and experience.

Role-play

The word 'play' has not been respected in recent years by critics of primary education. This may be because its implications have not been understood and it has been associated with what has been seen as an indulgent child-centred teaching style. Ironically, its use has been extended and refined as role-play in business and management education, as manoeuvres and tactical exercises in military training, and as simulation in a host of skill-demanding activities.

We need to re-establish confidence in the simulation processes of primary education from which our children are genetically programmed to learn. We know, from our experience as parents or teachers that children who are enabled to elaborate the technological and material aspects of their play grow up with the confidence to use them in adult life and to contribute to their development.

Play should not be thought of as an indulgent activity to be confined to the nursery and left behind as quickly as possible; we should take note of the willingness which children demonstrate to apply their energy to establishing rules of play. Children often appeal to adults for help in arbitration over the rules of a game. There is also often a need to revise the scenario and intensify the demands made by the story.

This playing out of scenarios is also a method used by professional designers whose jobs include seeing how a proposed new piece of hardware would affect the lives of its users, and of those with whom they work. Any serious approach to the design of a product must involve some attempt to imagine how it will be used. The more sophisticated the product, the more soundly based and accurate the attempt must be. Accurate scenarios or descriptions of potential users, their environments and their purposes are essential to design and these can only be achieved by observation of real-life users in real situations. Once such a scenario is created it can be updated or developed in the light of new observation. It serves to reveal design opportunities and provides a framework for generating and testing ideas in as close a simulation of reality as is possible.

The same theory can be applied to children; there is no good reason for insisting that first or second ideas must be completed. It is in the nature of children's play that most of the ideas that arise do so from very limited experience. Some are quickly seen to be impractical, others need to be explored in trial-play where failure is not serious and cost, in terms of investment or disappointment, is not high. This is why at this important stage, responsibility for ideas lies with the children or the designers and not the teacher or client. Teachers and parents, like good clients, need to support and understand the processes of design and develop trust in it.

An important function of play is to gain experience more rapidly and in greater safety than could be achieved by participating fully in adult life. It is essential to the health of our species that, in the course of their play, our young experience apprehensions and fluctuations in their confidence as they dare themselves to take physical and intellectual 'risks'.

NEW EXPERIENCES

A third natural purpose for which children use materials is to come to terms with a new or challenging experience. They are not, in this case, taking up a material to enjoy it, nor seeking to make something to fill out a play scenario. They take up the material for the purpose of drawing, painting or making a model about something which has impressed them and which they want to think more about; first for themselves and then possibly to share with someone else. Examples include a visit to a theatre, a holiday or an impressive encounter with a person or an animal, which needs to be recollected and understood in terms of a suitable material.

This kind of motivation might include a need to work out an idea of importance to them which may not be wholly accessible in words. Such ideas can vary from fantasy, to geometrical structures or 'inventions' intended for what they see as real situations. Success may depend upon the previous experience of the child with a suitable range of materials and their continued availability. The greater the familiarity a child has in handling alternative materials the more likely they are to form their ideas in terms of those materials. If so, there is every possibility that the idea will be realizable. If experience is very limited some frustration may arise. This is most often because the medium available, or chosen, is not easily related to the idea. If, for example, the experience about which the child is concerned is geometric and linear, the presence of paint but the absence of pencils may make a satisfactory resolution impossible. Similarly, if the experience is one of storm, winds and rain the presence of a pencil and the absence of paint may be equally frustrating. Where a child is normally confident, the parent or teacher often needs to do little more than to identify the cause of frustration and provide a more appropriate material.

Some ideas and experiences require three-dimensional materials and cannot be adequately thought about, initially, in two-dimensional terms. Some experiences may require structures, either static, or with moving parts. When this is so nothing is more likely to develop in young children the conviction that they 'can't draw' than a supply only of drawing materials. What they may really require, to avoid this unfortunate loss of confidence, is a material with which to quickly realize an idea which involves structural or spatial relationships.

It must also be said that children vary in the ways that they think. Some are much more likely to build or assemble objects than they are to draw or to paint. Of those who like to draw or paint, some show a predisposition to make free and rapid gestures, others have a more deliberate, watchful approach. Variations also occur within individuals, according to the job in hand, so that sustained personal preferences may not emerge clearly in children for some years. Understanding that these variations are natural can allay anxiety on the part of teachers and parents so that in their responses to what children do, differences can be welcomed.

MEETING CRITERIA

Play on its own does not, of course, constitute an education but it does reveal an in-built predispostion for learning which can guide our approach as teachers. The three kinds of motivation discussed above are deeply human and have long been effectively used by good teachers to turn individual instincts into a responsible awareness, by a class, of what is worthwhile, what can be planned and what can be made or done.

Education for 'design and technology capability', as required by the

National Curriculum, provides the kind of structure and progression which will enable more teachers to achieve those outcomes.

Perhaps we have a tendency to see things as defined by what we have come to think of as their opposites. For this reason, we sometimes see the mind as opposed to the body, art as opposed to science, the industrial as opposed to the academic, the material as opposed to the spiritual and so on. These are dangerous perceptions, since in each of these pairings, one is a function of the other. That we actually know this is evident in that, when we seek to judge other countries or earlier cultures, we do so by assessing the extent to which their artefacts, environments and systems reflect their highest values. Design is our means of applying that criterion to our own time, and technology can be our means of meeting it.

10 PROVIDING FOR PRIMARY EDUCATION

THE purposes which naturally motivate children to use materials often overlap and may be sustained, for quite long periods, either by children working on their own or by a whole group. The parallels, which were indicated in Chapter 9, with the activity and methods of professional designers are intended to show how work and play, arising from naturally motivated beginnings, can be intensified and enriched, by responsive support and informed teaching.

The attainment targets for design and technology capability in the National Curriculum, provide one framework within which this teaching can occur. This chapter looks at those targets, and considers how children's existing purposes may relate to them.

ATTAINMENT TARGET ONE

Attainment Target One is concerned with children learning how to identify needs and opportunities for design and technological activities through investigation of the contexts of home, school, recreation, community, business and industry.

The different pleasures that children find in certain materials first leads them to discover their properties and how they might be used. The pleasure motive may then be sustained by helping children to note differences in materials and their use in a variety of contexts. Short-term explorations interspersed with opportunities to discuss what they have learnt, enable children to share and extend their experiences more fully than if they were left to their own devices. On the other hand, time may be required for individual and concentrated work following such discussion.

All our ideas need to be structured in a language or in materials and usually in both. Each material has its own potential for being made into particular kinds of structures: children's ability to model ideas in the mind depends upon their having a repertoire of experiences with materials and provision must be made for them to gain a range of experiences. These should include opportunities to rehearse their ideas using alternative materials, in order to learn to think in different practical ways. Viable proposals for design cannot be drawn out of the air; they are generated and refined in the interplay between particular minds and the properties of different types of matter.

Exploration of their environment is a natural activity for children of primary-school age. In learning to make proposals for activity in design and technology, a first requirement in any given environment is to discover what is there. To do this effectively requires access to materials which can be used as media for enquiry. The first proposals for such activity from children are usually concerned with modelling and responding to what they see around them. Later they will speculate about and begin to imagine what things would be like if they were altered, mildly or dramatically. Progression through each of these stages should not be unduly hurried but enjoyed and thoroughly explored by the children. Familiarity with what exists and speculation about its causes and effects can contribute substantially to achieving this target.

ATTAINMENT TARGET TWO

Attainment Target Two is concerned with children learning to generate a design specification, explore a design proposal and develop it into a realistic, appropriate and achievable design.

Sensitive teaching towards this target needs to take into account the fact that young children are already busy making proposals about the nature of the world as they encounter it. In seeking to explain events they are eager to suggest explanations and to speculate about possible causes and effects. It is upon this willingness that teachers must build.

Proposals, once they exist in a material form, can be seen and responded to by the maker, other children, and by a parent or teacher. Any observer can ask for clarification and can support the development of a proposal once it has begun to take shape. Its existence allows its originator to retain ownership of it in a way which is not possible when ideas are merely talked about.

Telling and listening to stories and playing at being characters in them, may have the same function for children as 'scenarios' for designers. In their need for materials to support their play, children first demonstrate their ability to identify opportunities for design. Their observation of what occurs in a variety of contexts, (home, work, sport, travel) often of adults whom they wish to emulate, is a prime source for their story-making and play. Initial observation of this kind, for which children exhibit a propensity, may be sustained and reinforced by teaching and can lead to more elaborated play involving further design opportunities. These can vary from brief exercises aimed at heightening perception and increasing discrimination (check lists, sketches, colour notes in paint, shapes in clay, rubbings and so on) to more sustained simulations of complete environments, allowing for fuller interpretations of roles, which may require props and systems (vehicles, signs, luggage, equipment and uniforms).

The element of play, provided that it fires the imagination and successfully engages children, is an advantage in education, since it allows for controlled progression in the demands which can be made upon the abilities of the children, supported by teaching in response to their own ideas.

Teachers adjust their classroom environments as far as possible to meet their own priorities. Here a permanent corner has been set aside, in a space between classrooms, where individuals can concentrate, from time to time, upon independent research.

Once an outline plot has been agreed, imaginative extension is required and can be taken as far as circumstance and the teacher's purpose allows. Intellectual problems arise, and provide scope for further and more precise research about the environment or procedures which are being simulated. Demands for cooperative teamwork, or the development of individual expertise can be built in by children or teacher, all within the protective framework of pretence. Pretend solutions to, for example, technical problems are a useful 'holding' device until greater knowledge and skill are available. If children and their teachers had to depend constantly upon 'real-life' solutions, the scope for their preparation for adult life would be severely curtailed.

The intervention of an adult, whether to advise, inform or instruct, should be calculated to ensure that it enables children's participation to be greater. The occasions when adults solve problems for children should be few and occasions when children are able to go forward to solve problems for themselves should be maximized. Teachers often achieve this by helping children to discern the nature of their problem more clearly by discussion and skilful questioning.

A child may propose that a model of a ship should have a funnel mounted on its upper deck. There is a problem as to how it shall be attached. Depending upon the materials used and the nature of the original purpose for this ship, a range of solutions exists. What the teacher is called upon to design here is not the funnel but the teaching which will enable the child to make, develop, and test, a reasonable proposal.

Learning to teach is largely concerned with becoming clear about what the intended outcomes from teaching are.

ATTAINMENT TARGET THREE

Attainment Target Three is concerned with children learning to make artefacts, systems and environments, preparing and working to a plan or proposal and identifying, managing and using appropriate resources, including knowledge and processes.

The pleasure that children take in a material or a range of materials can extend to making artefacts, systems and environments. Planning to work in a particular material is possible only on a basis of experience of it, and the management of resources presupposes experience of drawing upon a range of such resources. From the outset of their education in reception classes, many children are encouraged to share in the allocation of resources and to understand the necessity for choice and sharing. They can, in other words, learn to participate with the teacher in planning their own experience of resources.

The need for props to support their activity provides many children with motivation for planning working methods and ways of achieving what is required. Play is greatly concerned with planning; with setting the limits and rules which are required to make their model environment and scenarios work.

Every child's need for materials, as a means of coming to terms with ideas and experiences, includes satisfying their curiosity about how things work. Their methods involve observation, speculation and invention. Each of these has a place in learning to understand, for example, how a digestive system works, how a sewing machine is constructed or how a television picture reaches the screen. The need to think through a personal idea or a system can initiate drawing or modelling with appropriate materials, with children feeding new information into the process as it is found to be necessary.

ATTAINMENT TARGET FOUR

Attainment Target Four is concerned with children learning to develop, communicate and act upon an evaluation of the processes, products and effects of their design and technological activities and those of others, including those from other times and cultures.

In learning to make discriminating judgements, it is not difficult to say that one likes A better than B or vice versa. However, since initial preferences are based upon a host of subjective responses, it is much more difficult to explain them. It is much more illuminating for children and more useful for them, to look for differences between A and B; that is, to see more of what is presented, before there is any question of preference.

Judgements, once made, tend to inhibit perception. Preferences, where they are based upon superficial observation, may be little better than prejudice when applied to matters of race, culture, design or technology.

Children do not need to be taught to have preferences but they do need to be encouraged to look for similarities and differences before making choices. Only then are they able to begin to consider the factors which contribute to their judgement.

Awareness of the different qualities of materials, and discrimination in their use, are part of all design evaluation. Teaching should therefore aim to enable children not only to say what they like and what they do not like, but to identify and describe differences. Objective aspects of form, including weight, size, shape, colour, texture, apparent temperature and humidity are all factors which affect children's preferences often without their awareness of what is influencing them. In manufactured objects such factors as novelty, familiarity, resemblance, and mimicry can be influential, in addition to all factors relating to efficient function.

The arch enemy of design is prejudice; that is, attachment to a preference unsupported by clear perception. By encouraging children to draw, model, write or otherwise describe factors in things they like, their perceptions of them can be made analytical and articulate. Children are then more able to understand, appreciate and cope with preferences which differ from their own.

What applies to objects also applies to technology and its effects. Children need to learn to examine these for themselves and discriminate between them, before they can make informed judgements. Learning to see what is before them when assessing products and learning to see what is going on when assessing systems, is a significant and continuing asset for children throughout their education.

Stories and scenarios developed by children, in consultation with parents or teachers, enable them to identify needs and opportunities for design, for generating ideas, planning and carrying them out and evaluating their effectiveness within the play.

CONCEPTUAL MODELS: Children's use of material as a way of coming to terms with aspects of their experience, which is the third purpose identified in Chapter 9, is initially the most personal and individual of starting points. All children have to build up their own conceptual models of the world as they experience it. They do this is in conversation, by describing their experiences, and in the various models and images in which they depict events, as they occur, or soon after. Young children may be seen to try out, in their own bodies, gestures, postures and facial expressions which they have observed in people and animals. One value, to them, of such imitation is feeling

Providing for primary education

what it is like to make that expression or gesture; important learning towards acquiring what Howard Gardner *(1983, p237 et seq)* calls personal intelligence.

In their conversations, parents and teachers are aware that in order to take in new facts and experiences, children sometimes have to make profound adjustments to their existing conceptual models. Children, for example, are naturally interested in the composition of their family and build up their concept of 'family' from all their experiences of it. If a child has formed the habit of calling their mother's best friend 'Auntie', the consequences of learning that an aunt is strictly the sister of a parent, may lead to a whole series of questions, about other 'real' and 'honorary' relationships in order to readjust the original 'model'.

Children whose existing model is inflexible or incapable of adjustment may have no option but to reject new information. This may not matter very much when the only consequence is refusal to accept, for a time, that a favourite aunt, is 'not really my Auntie'. However, it is not too difficult to think of examples where failures to remodel the concept of 'family' or 'friends' might be more serious. Children need to be given every encouragement to make such readjustments successfully in a way which makes sense to them. This may include help with an alternative vocabulary which may make possible a different way of seeing the 'family'.

Changing perceptions

Similarly, in the matter of visual perception this need to modify preconceptions is encountered. The child who has learnt that red and blue make purple may be in some difficulty when with particular pigments they make brown. Or, take the case of children who in their pictures paint tree trunks brown but, faced with a real tree, discover that it is not so. Some can take the new perceptions in their stride and loosen the 'rules' of their old perceptions in favour of their learning new ones. Others cannot do so immediately. The provision of alternative materials (for example, clay, pastels, wire) can enable children to come to terms with new perceptions by finding alternative structures and models for them.

It is a commonplace experience amongst teachers of art in further and higher education that the use of different materials as media requires different perceptions. In teaching landscape painting or engraving to advanced students previously accustomed to working in other media, part of what they have to learn is to perceive landscape in the terms of the medium which is new to them. In painting, this may mean seeing landscape as stretches of colour without intricacies of line. On the other hand, engraving with a fine tool might mean discovering continuous patterns of lines in the landscape, previously invisible but which now leap into view

Children must learn to live in a three-dimensional world, and work in three-dimensional materials is more important than was understood when most schools were built. A number of small rural schools have enabled children to benefit, from time to time, by allocating one hall for joint practical work.

Materials in design and technology

Schools are designed environments. Many were designed a long time ago to meet objectives which were very different from those now stated in the National Curriculum. Here, however, an imaginative plan includes space for a small garden and pool, which for urban children is a source of pleasure and a resource for learning.

because of the limitations of the medium.

The same applies to children who, having access to different media, can also be freed from habitual perceptions when these prove to be inhibiting. It is important for children to find confirmation of their perceptual proposals in images made by others. For this reason they should have access to many kinds of pictures: drawings, paintings, photographs, diagrams and models, which offer confirmation for many different ways of perceiving and depicting reality. Those children who are confident of their powers may wish to make copies of images or models made by accomplished adults, with a view to learning from them. They can gain information in this way about the content of the work and about the technology involved. Sometimes what is learnt in this way is relevant to their own work and can contribute further to their confidence and competence.

This is different from other less confident children who, when encountering difficulty fitting their new perceptions into old models, seek to avoid the effort by copying convincing images made by someone else. Think of a child who has acquired the model-in-mind of a yacht or an umbrella from any of the typical, two-dimensional images found in alphabet books or charts. Consider further the same child confronted by a real yacht or umbrella with a need to learn something about its shape or construction. Making a drawing of it would be a reasonable way to begin. What happens when they start to draw?

The first marks made on paper, by the child, if the real subject is well observed, will bear no relationship whatever to the old mental image of the boat or the yacht, which has hitherto served so well. It is at this point that insecurity occurs and the desire, by the less confident child, for 'something to copy' arises.

There is however a potential hazard here, in that most published pictures, whether drawings, paintings, photographs or diagrams, are sophisticated products which rarely reveal the complex processes which produced them. These include observation, using a range of inquiry and study procedures, preliminary ideas, sketches, mock-ups and trials, re-thinks, sidetracks, dead-ends and fresh starts. Successful engagement in these processes demands persistence, and this is not acquired by copying their outcomes. Children who wish to learn from adults need to be brought into contact with the authentic processes which artists, designers, engineers and craftworkers use, even if these have to be adapted sometimes to the limitations of the classroom.

AUTHENTIC PROCESSES

There are a number of ways in which this is achieved in school: they include varied and rich collections of natural – as well as designed and made – objects to provoke interest in aesthetic, functional and structural properties and as subjects for study. Collections may be school or class-based or made by individuals or groups so that children have a personal stake and a proprietorial responsibility for developing and updating them. Examples might include making collections of:

- things having the same nominal colour to reveal the range of hues which may be given that name
- cloud textures cut from different pictures and of similar textures from other sources
- shiny-curved surfaces and matt-curved surfaces, within which other qualities may be noted and further sub-set collections started
- yarns from threads to ropes, showing different structures and fibres and colours which may lead to observation of other qualities and the need for other sub sets
- items selected by any child appointed to be a 'world expert' on foot prints, food packs, power sources or other special interests
- personal selections of different objects having one secret quality in common; others in the group could try to identify and contribute.

A useful purpose can be met if the aim of the collections is to sharpen sensory awareness and discrimination. The elements of form are much more easily identified when they are seen, contrasted and compared as tangible, physical entities than when merely talked or thought about as verbal abstractions.

Objects
Objects are useful for dismantling and, when practicable, reassembling. They can also be used for identification of design and technological content in 'found' objects, and reading everyday objects for structural and design similarities and differences. Examples might include: clockwork models, toys, collapsible furniture, packages, clothing, footwear, any objects or pairs of objects affording simple deconstruction to see how they were made. Older children can treat such objects like archaeological

specimens and deduce what materials, design choices and technologies were required for their manufacture and what might be concluded about the culture which produced them.

Visits

Children can benefit from visits to and from artists, craftworkers, designers and technologists, as well as planned visits to workshops, studios, shops, galleries, study centres, sculpture parks and industrial sites and museums. Designers and craftworkers, including sometimes former pupils, may be invited to visit and work in school.

Art activities involve technological processes, as here where papier mâché accurately replicates the principles applied in reinforcing plastics.

Contacts

Contacts with specialist teachers and students from all stages of education including further and higher education would also be helpful to children. It is also a good idea to embark upon local research; ask local manufacturers and traders who designs their goods, products, merchandise, publicity and so on. Negotiations for contacts should start from a clear idea of how children and teachers hope to benefit and consideration of what might be of mutual interest.

TEACHERS AS DESIGNERS

All primary school teachers design, to a greater or lesser extent, the environment in which they teach. Similarly, they design many of the procedures they use and which they negotiate with the children. The extent to which children can share in designing these procedures or in designing the use of time, varies but where possible, they develop some responsibility for making designs work. They may also feel able to make suggestions about how designs might be modified in order to work better, and recognize that design initiatives can be shared.

Opportunities for design and technological activities exist across the curriculum. Reference to the Design Council Report *Design in Primary Education (1987)* and the INSET Video Training Packages *Change in Practice (1990)* and *Stories as Starting Points for Design and Technology (1991)*, provide a useful framework within which the cross-curricular nature of design-related activities can be explored.

Design awareness and knowledge of technology may vary from teacher to teacher. Design and technology appear in every tool, system or artefact which people use. All can extend awareness by making a brief study to identify what we know, from tailoring to plant propogation, from knitting to wheel changing, from hang gliding to loft insulation, from papier mâché puppets to fibre-glass boat building. Most people find that they have access to more technology than they had realized, before making such a study. Once a few familiar areas are explored for their technological content, they can contribute significantly to a teacher's confidence. Key questions in relating such familiar matter to children's learning might include: 'What underlying principles are involved and how can children be led to understand them?' and 'What activities, with which materials, might best extend their understanding?'

It is a characteristic of good practice in teaching that it respects the integrity of the mind of the learner. This means a recognition that children act on a basis of their existing perception and understanding. Good design education, indeed good education, means enabling them to extend both. Validation of their existing understanding and confirmation of their existing perceptions, as far as they go, allows the children to dare to extend them. The art of teaching and parenting has much to do with making the learner confident enough of their ability to learn and respond to the new challenges which teachers and parents must offer. Activities and learning in design and technology provide a particularly rich and diverse range of opportunities in which to present these challenges.

■ THE NECESSARY RANGE OF MATERIALS

MATERIAL INFLUENCES

THIS chapter is concerned with the need to identify and to provide for children, the range of materials which is logically necessary for the continued development of their language, for the acquisition of concepts which are essential for capability in design and technology, and also for intellectual development and cultural understanding.

Unlike those who set up our first national education system some 170 years ago, we are in a position to know how important technology has been to the evolution of human intelligence and how materials not only influence perception and shape the culture, but also contribute to the education of individuals.

Perhaps the greatest single influence upon our thinking has been the development of print technology. The great period of book-based learning existed alongside the massive extension of all the other material-based technologies which we associate with the industrial revolution. Every recent exploration of the physical world has been facilitated by book-learning while language and literature have been enriched and extended by such exploration.

Now, however, we are entering a mode of learning which is neither from experience of material nor from literature, but increasingly from video and computer screens. While these media can produce convincing images, they offer no stereoscopic, tactile or olfactory experience. Although they are extremely valuable images they do not provide the child with the full range of sensory references which is necessary for understanding and using language.

It is difficult to quantify the amount of information which derives from

The necessary range of materials

these screens, but the time which we spend attending to them suggests that, both for us and our children, they constitute a major source of information. They provide transient images, filled with information which, by the nature of the medium, is not easily recalled for critical contemplation. For this reason some children are provoked to read but many are not. Either way, the traditional balance between book-learning and personal hands-on experience of the material world is disturbed and may ultimately be profoundly changed.

We have seen how materials are necessary 'to make sense' of language. Neither televisual nor computer images provide the equivalent of full sensory experience. Indeed they too require that children have a sufficient range of sensory experience to be able to understand them. We should not be surprised that children with least opportunities for sensory acquaintance with the material world find it difficult to relate to what they read and have increasing difficulty in understanding, in life terms, much of what they see on the electronic screen.

The provision of materially-based learning is therefore too important to be left to chance. To make sense of an image means to draw from it perceptions and ideas which we can relate successfully to our own experiences of the world. If children are to learn to cope both with language and technology-based images, we need urgently to review the basis upon which experience with materials is provided in primary schools.

From Montessori onwards, there has been tacit recognition in primary education that young children need to have access to a range of materials. Most opportunities are offered in what have been called 'creative activities', or 'art and craft' but also in support of topic work, environmental studies, home studies and increasingly in science, mathematics and technology. The value of practical work has been reaffirmed in programmes of study for the National Curriculum, but without overall guidance about the range of materials needed to support such work.

Many schools state their intention to provide children with experience of as wide a range of materials as possible. However, since what is possible varies with both circumstance and conviction, such a statement does not provide sufficient guidance for parents, teachers or others who may be responsible for providing materials. Nor does it prevent the provision of too many materials at any one time, which may result in much valuable

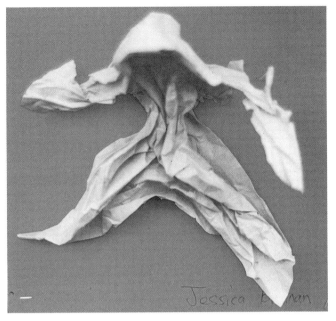

A teacher can sometimes accelerate learning by appearing to slow down a process to ensure that it is properly valued. Here a headteacher noticed the aptness of this running figure made by a child from a rapidly manipulated hand-towel. The teacher's response was not merely to admire it but to challenge every individual in the class to make equally enterprising use of the same material. The short-term result was an exhibition of their many ideas. The long-term result, from many such responses by the teacher, is a small school in which children's ideas are appreciated, shared and challenged.

experience being lost and material wasted. If education is designed to enable children to come to grips with the concepts of their time, the materials required are those which cover the essential range of experience.

In Chapter 6, we looked at aboriginal materials in general categories and discussed some of the concepts and structures which we have derived from their development and use. Since children need to be equipped to understand the principles underlying all structures they need to have experience, if not of all materials, then of all categories of material. Similarly, since experience is needed to furnish the means of modelling in the mind, it is necessary for children to become familiar with materials representing all categories.

ACCESS TO MATERIALS

In addition to the fundamental reasons stated above for a rational provision of materials, there are other more circumstantial reasons for urging that increased attention should be paid to the importance of materials in primary education.

The number of opportunities for children to have access to materials and to see skilled adults working with them as an integral part of their daily life, has steadily reduced over the last 170 years. A number of general reasons for this has been given in early chapters. The following specific reasons can be identified in different locations and circumstances. In spite of compensating opportunities which may exist for some children, the following changes contribute to the loss of opportunity for many.

The increase in the size of urban and suburban communities reduces the exposure of many children to mixed town and country life where work on the land and with materials is most evident. Increased traffic and other environmental hazards have reduced areas of access for many children. Greater specialization in agriculture and manufacturing industries means less variety of activity in any one area. Some have no small workshops in which the labours of craftworkers like blacksmiths, woodturners, tailors, dressmakers and coachbuilders might be familiar. Other areas have an excess of so-called 'craft shops' where the last thing one sees is material being genuinely worked. The exceptional efforts to bring craftworkers into school are most valuable but they cannot in themselves compensate for this loss.

The use of sales and reception staff in industry and the motor trade, designed to exclude customers from workshops, means that children do not see craftworkers on a day-to-day basis and do not have opportunities to admire skills in action.

Children may grow up unaware of the effort and job satisfactions associated with the professional use of materials and of what is required to maintain the fabric of their environment. One might pertinently ask how much the ubiquitous small workshops in Italy contribute to the emergence of so many good young designers in that country.

The automation of processes, tools and controls means a distancing of the worker from the material. The many advantages are gained at the loss to children of familiarity with the material itself.

Pre-fabrication in the factory behind closed doors reduces opportunities to see craftpersons at work on site and reduces the need for itinerant workers with maintenance and repair skills. Factories now tend to hide behind their façade and image where previously their goods and services were more visible.

For all of these reasons, we cannot assume that most children have dug soil, sieved sand, diverted or bridged a stream, bowled a hoop, built a shelter, made a costume, climbed a tree, built a wall, kneaded bread, or balanced a seesaw. None of these activities is, in itself, indispensable but each represents a basic experience which needs to be paralleled or replaced by something similar.

Using local materials
Although the categories set out below are the same as those which were used to classify 'aboriginal' materials, there is no suggestion that children need to handle *natural* materials in all categories. Rather it is suggested that the provision of some materials, manufactured or natural, from each of the categories provides a basic repertoire of experiences, without which children's access to certain concepts and related language might be seriously impaired.

It is acknowledged that children can benefit from finding the location and sources of materials and working with some in their natural form. Working with a raw material, such as clay or wool, is a valuable way for children to learn how the properties of materials may be processed and adapted by machinery, prior to being taken through design and technology, to manufacture and marketing.

Such opportunities vary from region to region but may together contribute to the 'national' curriculum a better understanding of materials. Depending upon the age and capacity of children, finding and using raw materials can lead to exploration of ways in which it is used locally. Social and industrial history may reveal how related industries emerged. Local, natural materials often also have potential for teaching physical

Human life is conducted within a balance of adventure and security. Any new environment is potentially threatening and, as we explore it, we identify 'islands' of security within it. The rock is a place of safe congregation for these special-school children while they gather courage to venture further into the sea. Features of landscape, like natural amphitheatres and terraces have long been seen as congregation places.

science and for art and design. The different but complementary natures of contributing subjects are often best revealed in such local studies.

As a general principle it is important to recognize the relative convenience of sophisticated ready-made materials against the educational values which can be derived from using more basic materials. The choice is a matter of the teacher's experience and judgement.

Local industries and businesses are often happy to supply materials, either cheap at the point of production or from waste or offcuts which can amount to a big saving in some categories and allow for greater expenditure in others. However, advice should be taken to ensure that there are no unforseen hazards in the use of such material. Many apparently harmless materials can be hazardous in the hands of young children.

RODS AND LEVERS

Rods and tubes are now pre-packed in the form of 'straws', round or square-sectioned dowels in jelutong, balsawood, plastic or alloy. There are also many kinds of natural sticks which can simulate, in miniature, authentic ways of building. Card and paper can work for two-dimensional models but they need to be folded or rolled into tubes in order to function in three dimensions.

There are a number of easy-to-use ways of assembling rectangular constructions and wheeled models. These have their uses where they build on previous learning but are not appropriate where they by-pass fundamentals. Teaching should include children's investigation of ways of construction using fixed and movable joints, in order to encounter genuine technological problems. Some children demonstrate ingenuity and insight which, in sharing problems and solutions, can generate further teaching. For this reason the ready-made solution is rarely the one from which most can be learnt.

Frameworks often provide a skeleton shape which is fleshed out or clothed in other materials. Alternative means of building bulky forms may include boxes, stuffed or inflated bags, or wire cages which can be later removed or left in position depending upon the nature of the covering material. Having devised a process, children are glad to find examples of the same principles applied in industry and more sophisticated technology.

Three-dimensional models often involve moving parts and mechanical functions. Much can be learnt from the use of hand-generated power so that children can test the effects of leverage and gearing. They may explore the use of water, air, gravity, fuels or the much exploited elastic band, to provide or store energy. These explorations, with appropriate teaching, may yield more learning in design and technology, at much less expense, than a ready application of commercially produced power sources or kits.

Strategies used in teaching can vary and lead to different outcomes and benefits. A range of materials can provoke a number of possible ways of working. Alternatively, concentration upon the possibilities of a single material can be valuable: while it may limit what it is possible to make, the learning can sometimes be more focused and the sharing of what is learnt more precise. Both strategies are useful and can be adopted as required. Success may depend upon there being a sufficient quantity of any one material to complete a project.

GASES AND FLUIDS

Gases and fluids are still readily represented by the life-sustaining media, air and water. Water is a traditional material in nursery and infant education but is worth fuller exploration by older children as a way into understanding flow, volume, power and energy conversion, conservation, efficiency and the behaviour of fluids by observation. Designing and making boats, hot-air balloons and gliders, or devising ways of harnessing the power of wind and water, present opportunities for learning principles and for extending and enriching language. Much material- and sense-based learning can be introduced by a simple act like taking a large sheet of paper, fabric and hardboard, respectively, across a playground on a windy day and noting the responses of the materials.

Other liquids with observable useful properties include household oils and modern detergents.

DYES AND PIGMENTS

Pigments now come in block, powder or liquid forms and vary considerably in quality. Some pigments, the so-called earth colours for example, are easily obtained and cheap, while other, rarer ones are expensive. Cheap block paints contain fillers and binders and often include cheaper dyes as colouring matter in place of the more expensive pigments. Good quality powder paints generally give better colour potency for the money. The convenience of prepared liquid paints makes them attractive but, on their own, they offer a more restricted experience for children. Using powder, water and a little PVA can allow teaching and learning to occur by observation and discussion about colours and qualities in paint.

Some qualities, particularly of colour purity and brilliance are unobtainable in cheap pigments.

Different densities and viscosities are required for making transparent or opaque paintings and for printing by block or screen methods. All such explorations afford opportunities for learning relevant technology and applying it creatively in art or design.

Cheap but strong dyes are available for making inks and for painting or printing on fabrics, which are quite adequate for short-term use. Colourfast dyes or fabric printing inks, which are more expensive, are essential only when their special properties become relevant.

Opportunities for powerful and skilful use of pigments and dyes occur in the visual and performing arts, but in other contexts children are often very aware of colour, without realizing how it functions. Examples include food, the interior and exterior surfaces of buildings, vehicles, specialized clothing, tools and equipment.

The colour of light varies according to sources, seasons, and times of day, and influences our perception of the colour of applied dyes and pigments. Children should observe differences, mix colours and note and discuss results. Without the basis of practical experience, so-called colour theory remains unstructured in children's minds and is easily forgotten.

TOOLS, INSTRUMENTS AND SMALL UNITS

Some tools and instruments can be invented and made by children when measurements are required or where unaided hands and eyes prove to be inadequate. Learning to use basic tools is important.

The small units which are supplied for counting and sorting and developing mathematical ideas are of standard form. The tendency of children to put them to other uses, often divergent and imaginative, shows how important such materials are to them. They need to have an ample supply of small units, which need not be finished to mathematical regularity.

Some sets of small units of similar size and shape are also useful. They may range through such items as seeds, beads, cubes, short rods, shells, dominoes or other tokens. The guiding principle is that there should be enough of each set of such units for them to be used in quantity like tesserae or bricks.

Other assortments may involve qualities other than shape and size, presented in similar or mixed sets. Children will use them for designing and assembling models and for constructing, adjusting and remaking flow diagrams, systems and desk-top simulations of everything from road and rail traffic to sea and land battles, from town plans to market layouts. When making all such conceptual models, small, inexpensive units of similar scale are indispensable.

Adequate quantities of similar units are required and can be built up as they are seen to be useful. The idea that small units assemble into large or long arrangements is at the core of mental models for atomic and molecular structures, social groupings and language. Children who do not have experience of such modelling may have difficulty in grasping and applying such abstractions.

Quantity may be important; for example a 'load of bricks' in miniature is quite a different proposition from half-a-dozen bricks, irrespective of size. Similarly a bale of a single yarn offers different possibilities from a number of hanks or skeins of different yarns.

The term 'junk modelling' was rejected in Chapter 6 as a demeaning description of an essential activity for children. Found or salvaged materials, often quite small units, such as boxes, cartons, and tubes, can be assembled to stand for or represent whatever is required to support a current class project. More can usefully be done to ensure progression in such work by teaching children to observe the qualities of materials and to relate them to the qualities of what is being modelled.

Consideration should be given to the availability of materials from more of the categories identified here to extend the range and quality of the work. Found materials themselves may be usefully collected and sorted into these or similar categories as children come to understand them.

Far from being 'junk', a range of materials to use as a basis for modelling is amongst the most important provision in a school. That much of it is derived from 'waste' does not matter if it provides an adequate range of alternative ways of constructing and modelling ideas.

A group of children, seated round a table on which a range of materials is provocatively displayed, will not be short of ideas as to what might be made from them. After some exchange of the many ideas which the materials provoke, individuals may be invited to take to their workplace a selection of the materials which they expect to use to realize their idea. It is important that the teacher or parent understands enough of what is intended, so that he or she can ensure a sufficient supply of the chosen materials. Support and advice may be needed when difficulties are encountered. However, since the children's ideas are a direct response to the actual range of materials present, such difficulties are unlikely to be insurmountable.

Disparate objects will occur, for example a boat, a bird, a house, a horse, a soldier, an aeroplane, each with characteristics which can further fire the imagination. Although they may not be of the same scale, such an assembly of objects and characters, conceived and made by the children, can quickly be woven into a story of their devising. What follows, in developing the story or integrating the elements into a more elaborated model, designing a technological development or undertaking further related study, is a matter for the teacher and may provide an opportunity for negotiated organization.

The range of materials required for such a start might include: wire, paper, string, adhesives, cartons, boxes, wood strips, rags or textile waste,

Weaving is a technology which has been fundamental to human development: by discovering how to turn yarns into fabrics our ancestors liberated themselves from many constraints. Weaving is capable of almost unlimited sophistication and has a vast range of applications. Textiles constitute a technology which is so central to our way of life that we tend to overlook its potential for education in technological capability.

clay, card offcuts, paper sheets, paint, larger pieces of fabric, yarns and thread, along with appropriate basic tools.

This example of a successful approach is but one of many. However, it is one which any experienced teacher, with a confident class, can test in practice. It serves to demonstrate the value of having available, from time to time, a spread of materials from most categories, which can provide choice and potential. Very often the responses of individuals will reflect prevailing interests of the class, and may therefore tie in with other work in a particular aspect of the curriculum.

STRINGS

Strings present no difficulty in supply and, for that reason, their importance may be overlooked though their usefulness, together with their linear metaphors, can hardly be exaggerated.

We string many ideas together which only work because they are tied to each other in that way. We string letters together to make words and sentences which we can record on magnetic tapes or as strings of data.

Strings, chains and lines are used for making measurements and when run over pulleys or cogwheels at constant speed can measure time. Although natural fibres are still used, synthetic fibres and filaments can be made which have different properties. Continuous filament does not have to be twisted together to form a continuous string whereas short fibres are spun into yarns and further twisted into strings and ropes.

Belt drives are used to transfer motion from one axle to another and when moved from one pulley to another of different diameter, will change speed and mechanical efficiency. Strings of metal filament are used to convey electric current. Optical fibres can convey light around corners. Belt drives can be used for linking motors and also, with only slightly more sophistication, for timing and synchronizing mechanical functions.

Webs and circuits require linear thinking and modern plastic and adhesive tapes make their constriction and modification easy and highly visible for display and communication.

SHEETS

Sheets are everywhere, and can now include card, metal and plastics as well as textiles. They can be of manufactured or natural fibre, opaque, translucent or transparent, rigid or flexible, plain or coloured, providing a splendid range of properties for exploration and use.

In addition to a whole range of textile and artistic skills including drawing, painting and collage, modern uses and skills include printing from flat and relief surfaces, and photographic printing.

Techniques involved may include: cutting, shaping, laminating, folding, fitting, card modelling and models for sheet-metal construction, framing, mounting, projecting as in map making, and slide or film images.

The necessary range of materials

Particular applications, to which sheet materials can bring inspiration include solid geometry, the study of light and lighting, interior and theatre design and architecture. Also textile constructions in clothing, travel and sports equipment, and powered aircraft and gliders.

PLASTICS AND ADHESIVES

The basic material for experiencing plasticity and additive modelling is clay which, with plaster, allows for the experience of moulding and casting. It is such a fundamental means of thinking three-dimensionally that it deserves priority. Substitutes are inadequate for primary modelling experience but some of the variety of modern plastic materials which are available, are worth exploring for their different properties and potentials.

Versatile, PVA-based glues are water miscible when wet and can be removed with methylated spirits if dry. Latex adhesive can be removed readily from smooth surfaces including paper but a solvent is needed if absorbed into fabric. Most adhesives require some practice to use efficiently. Unfamiliar, volatile, instant or hot adhesives are not appropriate for unsupervised use by young children.

CONTAINERS

Containers can vary, from thimble size to packing cases large enough to act as a small room, and are made from materials with very different properties, such as rigid, flexible, elastic, transparent, waterproof, strong and so on. Each of these properties can be exploited in different ways.

Apart from their usefulness for storage and as readily adapted interiors and mini-environments, they may provide building material, and assist in teaching about air, fluids, volume and sound where they can assist in supplying pneumatic or hydraulic energy and controls. Some containers are well adapted for making and displaying collections.

LARGE UNITS

Large units other than natural objects and materials can be contrived and can include such varied things as indoor or outdoor places and areas for development, locations for particular activity, outdoor apparatus, pools and troughs, large plastic blocks for assembling or building, real-life building materials, soft insulating blocks and bricks for carving. Outdoor and indoor apparatus can be adapted for short-term occupation. So called 'spare' rooms, balconies and covered areas, ends of corridors and covered outdoor areas often allow for quality work not always possible within a classroom. These areas all need to be valued for design and technology and creative art activity.

There is special aesthetic value in children handling materials which involve whole-body responses and controls. Environments requiring different qualities of effort at different levels may invoke ideas which involve sound, light, voice and movement.

FOOD

It is inconceivable that any material provides greater benefit or makes greater demands upon us than do foodstuffs. No material so modifies and controls our behaviour and attitudes. Because it is a daily part of our lives its potential for design and technology teaching is easily overlooked. Foodstuffs are used to construct confections ranging from the modest to the architectural. Food is coloured either naturally as it comes to us or in the processes of cooking or by the addition of edible colour. It is presented in all scales and quantities about which design judgements must be made. Now it can can be extruded, like sausages and some cheeses, meats and pastas or rolled and processed. Its provision, distribution, preparation and disposal all needs to be planned. Considered as a material it has a unique place in our lives and our education.

SAFETY AND DANGER

The three-dimensional world is dangerous. The safety precautions, which parents and teachers adopt in environments used by children should not conceal from them the fact of danger. Rules of appropriate behaviour need to be balanced by teaching which provides opportunities for children to progressively recognize sources of danger. In spite of precautions and sensible design, any real environment is potentially dangerous and can be said to be safe only as far as the sources of danger are understood by the people who use it.

As part of the dangerous, three-dimensional world, technology itself is never free of hazard. Technology allows us to put materials and energy to work in ways which are safe only to the extent that the behaviour and effects of the forces involved are understood. These are best understood through practical involvement with them as part of one's education. In the world at large, responsibility for understanding has to be delegated, to some extent, to experts but we all need to know when to withdraw or to intervene appropriately when experts let us down or are not available.

MARKETING

Children need to learn how manufactured goods reach the customer. What is taught should meet the children's need to know and ability to understand. They should learn to understand the difference between prices and values and meanings of words like 'costs', 'profit', 'needs' and 'wants'. Consideration should be given to planning how children can come to an understanding that goods or services for sale should have real value.

Parents and teachers may value a child's work far beyond its present market worth, simply because it is indicative of emerging ability. Such work should be cherished in the school or family. To sell such work at less than its value would be to diminish it, just as to sell it for more than its market value would be patronizing. A worse practice is to involve children in making cheap imitations of market goods which are purchased by parents and friends as an act of charity. It is better to raise money by other means and to postpone sales of children's work until they have acquired the maturity and skill which are needed to make and present work which is of marketable value. The timing of that achievement can vary and depends upon what and how things are made. Technology in the form of photocopiers, printers and word processors make some objectives achievable earlier than others, even for children.

Children live in a three-dimensional world and must learn to share in responsibility for it. The necessary range of materials for primary education must be that which enables children to experience and test the main ideas and structures which they inherit. Such provision cannot be left to chance nor does it mean that children must handle or use all possible materials. We need to give priority to those materials which provide the essential experiences.

GENERAL COMMENT

In future, it may be necessary to agree about the range of materials to which children should have access; to draw up what might be thought of as the curriculum of matter. In the meantime, what is suggested here is a relatively simple rationale, for offering mainly inexpensive modern materials, which together cover the main characteristic properties and functions that were found in the materials from which our technologies and cultures have been constructed and upon which our ideas have been modelled.

12 FURTHER STEPS FOR EDUCATION

A sustained look at materials and the ways that we use them makes us aware of the evolutionary nature of technology and perhaps able to see education in a similar light.

EVOLUTION: Technology evolves by developing new ways of using material resources, and by modifying what has been done before. Each new technology exists alongside the old, for a time, and if it proves to be merely novel it does not thrive. If, however, it meets general requirements better than the old, it soon gains ascendancy over it. The old does not entirely disappear however, and in some tiny niche, from which it draws strength, it can survive and reappear should circumstances ever change again in its favour.

However, since we are not entirely passive in these matters, we can intervene and, by design, look ahead to anticipate change and specify the technology most likely to meet new needs and purposes. Clarifying purposes and making decisions calculated to meet them, is a function of forward-looking design.

Education also evolves. Neither the policies which describe its purposes, nor the teaching which aims to achieve them, can remain static. Choices are made on the basis of existing knowledge and perceptions, but with future requirements in mind. However, since success in education is not measured by intention but by what children actually know, understand and are able to do, teachers and policy-makers need to be able, like designers, to modify their choices in the light of what actually happens.

Even if the targets for education were to remain the same, the means of achieving them would vary, because children vary. This is why the non-

Further steps for education

statutory guidance which accompanies the National Curriculum does not specify the method of delivery. Pupils change from one year to the next and from one class to another, not least because change in them is demanded by changes in the world in which they have to succeed. Because of the increase in international, economic and technological planning, changes increase in both scale and rapidity. Children sense the physical, economic, and social environments to which they have to adapt, although they need our help to understand and function in them. Education for change and adaptability includes education for capability in design and technology.

While some things in education change, two fundamental demands which survival makes upon any culture do not: the first is to ensure that its wisdom, knowledge and skills are passed on to the following generation; the second is to ensure that the young are raised to be willing and able to accept responsibility for the future maintenance and adaptation of the culture.

The successful evolution of our species has determined that our young are energetic and learn through activity. Nature has ordained that they learn from their play. They could do this without the cooperation of adults, though what they learnt might be very unfortunate and would, in any case, have more to do with their personal survival than with the continuance of our culture.

Evolution has determined, however, that human adults are involved by making sure that our young remain dependent upon us for a good many years. This useful strategy enables adults to have command over what is taught, if not entirely over what is learnt. In order to take advantage of this opportunity, teaching must be designed to ensure that it goes far beyond immediate needs, and goes on to include the knowledge and skills which are vital for the survival and development of the culture.

During their extended period of living with us and imitating our ways, the young are able to see opportunities for improving upon our methods and lifestyles. This too is to the advantage of the culture, since the better we educate the young the more they are likely to wish to improve upon previous achievements. Whether their perceptions are accurate and

This playground apparatus of stripped logs provides an excellent environment for exploration at many levels. The logs are placed side by side to create horizontal, vertical and inclined planes, which act as floors, walls and ramps. More open structures function as ladders, fences, windows and doorways. Diagonal poles act as stays, which buttress the higher parts of the structure. The material offers a challenge to courage and intellect.

Materials in design and technology

Conventional images are useful symbols but contain so little reference to reality that they can interfere with the development of confidence in drawing.

whether they are capable of acting wisely upon them will reflect the effectiveness of our teaching. The National Curriculum acknowledges the importance of children learning to see, for themselves, opportunities for design and technology activity, by giving it pride of place in the first attainment target for the subject. It aims equally, however, to ensure that they learn to act wisely upon their perceptions, by being taught to plan, carry out and evaluate design proposals, as made clear in targets two, three and four, respectively.

Most parents admit to being made to reassess their own values and behaviour under the critical gaze of their offspring. However, this two-way process, of mutual regard and assessment, is easily overlooked when we design our education systems. Certainly English literature has been enriched by the many writers who have recalled the way they saw the manners and modes of adults who figured in their childhood. We cannot assume that critical perceptions are confined to the few children who go on to achieve literary success. Indeed we have only to observe children at play to see how knowing they are about adult behaviour. So strong are the roots of observation and imitation that we are sometimes exasperated to find that children know far more about our mannerisms than about the matter we intended to teach them. Awareness of the independent activity of children which occurs across the interface between children and teachers is a factor to be aware of when we design our meeting ground.

TEACHERS AS DESIGNERS

All adults involved in educating children – whether as parents, teachers or policy-makers – assume some of the characteristics of the designer. The design of a National Curriculum seeks to determine what is taught and, like all design, we should expect it to be modified and refined in the light of experience. No matter how well-balanced and crafted it is in theory, a curriculum is effective, in practice, only to the extent that children actually acquire the knowledge and develop the skills which are intended. The task of matching the content of a curriculum to particular groups of children is the most neglected aspect of curriculum design despite the fact that it may well be the most important. The erroneous idea that a product is well enough designed if it is well engineered is one which has resulted in many goods remaining unsold.

Designers working in the information technology industry, which makes products which help people to do things, have learnt that, in addition to the engineering aspects of design, it is also necessary to ensure that a product matches the way intended users think about what they do. They have learnt this the hard way by seeing that competitors who take the trouble to attend to this aspect of design lead the market.

It would be a pity if our well-engineered National Curriculum were to remain on the shelf because the importance of matching it to particular children is underestimated. Achieving this match, by translating curriculum content into teaching strategies and into appropriate interaction with particular children calls upon the highest professional skills of teachers. However, because these skills are manifest in the way teachers and children relate to each other, it is all too easy to regard them as aspects of character and personality of the good teacher. For as long as we regard these vital professional skills in this way we shall not train our young teachers properly nor make appropriate use of them.

It is true that effective teachers respect their pupils and have an interest in what they think, but genuine interest and respect come, not so much from teachers' personalities, as from their knowledge of children. Such knowledge is gained by observing them at work and learning how their perceptions influence their behaviour and their learning. Declared 'interest' and 'respect' which are not based upon a knowledge of children's thinking soon degenerate into frustration, hollow sentiment, or both. If we are to train teachers well we must answer precise questions about how the best practitioners in primary teaching first come to know and then always to respect the integrity which they discover in the minds of their pupils.

This photograph, taken through a Fifth Avenue shop window, shows how, in this most sophisticated urban environment, the appeal of a dress is related, knowingly, to the natural materials upon which aesthetic judgements are formed.

Many of the best teachers can give clear accounts of how they have learnt the importance of their pupils' perceptions. One such teacher said, 'Do you not think that we tend to regard as unintelligent those pupils who simply order their experience in ways which are different from the way we order ours?' What teachers of this calibre demonstrate is not that they abandon their perceptions in favour of those of their pupils but rather, that by taking the trouble to understand their pupils' ways of thinking, they become able to teach them.

Many recruits to teaching are genuinely motivated and bring high ideals to their initial training, but some of them fail because they do not learn to 'read' the thought structures used by their pupils. Teachers and parents who involve their children in practical work with materials give

Materials in design and technology

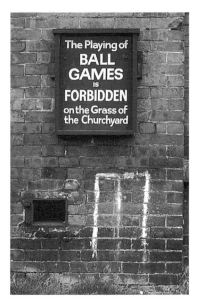

Unresolved priorities, like those of adult and child, are reflected in physical and environmental confusion.

themselves opportunities to learn how they both see and think. They learn also that, within their necessarily limited experience and perception, children behave in ways that make sense to them. Teachers who gain that knowledge can both respect the models which inform their pupils' thinking and challenge them precisely. This they achieve by designing lessons and activities which relate to pupils' ways of thinking, but require them to extend their thought and to acquire new models.

Some of the methods teachers use, in the design of teaching strategies, seem to be supported by those increasingly used in advanced fields of design. The need for user-friendliness in hardware (machines and equipment) and software (programmes) has led to a new awareness of the importance of human factors in design. The information which is held or produced by computers and electronic instruments must be presented in the way which is most appropriate for the use which is to be made of it. The interface, that is the keyboard or other means, which is used to communicate with the computer, must be designed to meet the thought processes and workstyles of the intended users. It is necessary, not only to design the logic and mechanics correctly, so that it works within itself, it is necessary also to ensure that the design functions in a way which actually 'fits' the thought processes of the intended user.

THE TEACHER AS INTERFACE DESIGNER

Teachers need to design teaching methods so that they are well structured but also so that the content of the teaching is accessible and usable by the learner. Successful teaching is based upon knowledge of the subject-in-hand and also upon knowledge of the child. Those parents and teachers, who achieve this, do so by observing their children at work and play, in much the same way that designers have learnt to observe the intended users of their products.

It has taken a long time to learn the importance of design in industry and we have been taught to do so by the success of our competitors. We now know that the higher education of designers must include preparing them for interaction design; that is to prepare the ground between the well-engineered product and the way the user will react with it.

It must not take us as long to learn how to use teachers. Our competitors may not give us a second chance. Teachers must be trained to prepare the ground between the well-engineered curriculum and the children whom we intend to use it. If quality is our objective, and we need not bother to look for economic success without it, such training is as necessary and achievable in eduction as it is in design.

The cost of educating the nation's children competes with other demands for resources. Confidence in education, as a long-term investment, is compared with others which may seem to yield more immediate gains. Education has not been seen to be successful for all children or even, in the eyes of some critics, for the majority. Too many children emerge from the education system with a sense that it has been of little value to them. The achievement of confidence in education, as an investment, requires that more children are able to identify and to demonstrate the success of their education in their own terms and in those of the community whom they serve.

Teachers, like designers, need to be able to draw upon more than their own knowledge and from more subjects than they can fully command by themselves. The more the skills of teachers are seen as professional and requiring proper training, the more it becomes evident that they need to be supported and supplemented by disciplines other than their own. More people, in addition to fully-trained teachers, need to be involved in education, particularly to assist children in relating their knowledge and skills to the world beyond school. Even with the vastly increased resources which technology can bring to the service of education, teachers on their own cannot adequately represent the opportunities and challenges which exist in the world for children.

CHILDREN AS DESIGNERS

A particular function of materials in education, is the means they provide for children to organize their responses and thoughts about that world, at each stage of their development. This they can do in ways which make sense, first to them and, subsequently, to others. By providing scope for action by children, materials enable them to make their thinking evident.

The inclusion of design and technology and art in the National Curriculum for all children is recognition that it makes no sense to postpone opportunities for practical action until children acquire adult status. Practical activity, with materials, provides the best opportunities for children, and others, to see how their capacity to act is progressing towards meeting the kinds of demands which adult life will make upon them. The proposals that children make, for design and practical development, provide the best opportunity, for parents and teachers, to demonstrate the value of investment in their education.

In order that the true potential of children can be identified, it is important that they are encouraged to work at their own levels of understanding and not merely to carry out the instructions of adults who know what to do. By seeing how children tackle problems one can learn to 'read' how they think, and what they do and do not grasp. One can see how they relate a cylinder to a plane or curved surface, how they attempt to attach one to another, and what they do about the difficulties they encounter. One can assess their capacity for making judgements and their persistence: one can teach by helping them to see the components of their problem differently and to see alternative possibilities. It is unusual for problems, encountered in activity with materials, to be unique to one child. The difficulty encountered by one child, at one moment, is likely to be encountered by another very soon. Because of the practical context within which they are working, children see the sense in sharing knowledge gained and the insights acquired in solving problems. They benefit from learning to listen to, and to evaluate, each other's propositions.

Any child's proposition which arises in a practical situation, contains evidence of his or her perceptions and may include understanding or misunderstanding which the teacher, skilled in observation, can pick up and use for further teaching with a group or an individual.

Competition, in terms of ideas, is indeed a part of economic life. Competition within education has most value when the assessment of ideas, products, or designs, is made, not merely to identify a winner but in order to see what is different, in technical or human terms, about the models which 'perform' best. Full educational value requires also that less successful models are scrutinized for the ideas which they contain and to discover how and why they were not more successfully implemented. Relative failures in design often provide the very insight which takes the designer forward to a new position of advantage for the next problem.

Discrimination is needed in assessing performance. Some models, made by children – or more accurately, some aspects of models – are intended to function practically, for example wheels that turn or boats that float and so on. Some modelled parts, however, are intended to function only to represent or stand for something which works. Parents and teachers need to understand this necessary aspect of modelling. It is a legitimate aspect which is extensively used by engineers and designers. The 'mock-up' is a very useful way of setting an idea into a real context to see how it will relate to objects, systems and people who pretend to use it, without the expense of actually making it work. Similarly, in drawing up a procedure or programme, a designer may simply draw a box and label it 'control unit' or 'distributor', without having the least notion, at that stage, of how it will be made to function. At a later stage in the design, this unit may have disappeared or, if retained, may need to be developed in detail.

This principle makes good practice in education. Children's ideas, although vehemently expressed, are transitory. To the extent that they are based upon their current perceptions, most proposals or ideas need to be modified and some replaced, as opportunities for fuller perceptions occur. Even if our purpose were exclusively to develop engineers who could make things which work, we should not serve our best interest or theirs by denying them opportunities to dream up solutions which are, for them at present, impossible to realize.

It is impossible for children to make sense of the world without the use of media. Gesture, drawing, painting, modelling, speech, writing, music, mime, dance, drama, signing, signalling, designing and constructing all involve the manipulation of properties of matter. Information comes to us through our senses, but concepts do not come ready-made. These we have to construct for ourselves, but we cannot do so without sensory and manipulatory access to media. The formation of concepts, or doing something in response to sense-borne information, demands the use of materially-structured language and media. Those who teach very young children and those who have learning difficulties are sometimes quicker to recognize this fact than teachers of children who learn more readily. Successful primary teaching has demonstrated the value of design-related work in all areas of the curriculum.

A wholesome future lies ahead for education as it extends this work through its new experience of teaching design and technology. By placing its emphasis on developing a capability for it, in all children, the subject lends support and national purpose to the best of existing practice in primary education. By drawing attention to the need for continuity and progress, through all the school years, the curriculum for design and technology provides an impetus toward a fuller realization of the educational importance of work with materials. If this continuity and progress is achieved it must, inevitably, lead children to a richer appreciation of human enterprise and a greater readiness to accept responsibility for its future.

Materials generate their own discipline which is imposed not by authority but by possibilities which, once seen, demand to be realized.

13 CONCLUSION

TECHNOLOGY is older than humankind and is founded upon the characteristic limitations and potential of found materials. Exploration and use of materials, over countless generations, have contributed to the evolution of human intelligence and continue to exert enormous influence upon the ways in which we think. In modern societies technology is important not only to the economy but also to the development of ideas and to the renewal of language. Access to the abstractions of language is by means of the models used to construct them.

Children learn about the world through senses tuned to material changes and they must therefore engage in the risky business of making and testing their own propositions about what they encounter. They do this by using any available material to model their ideas. The roles of teacher and parent include understanding the significance of this activity and facilitating the refinement of children's propositions in the light of their experience.

The National Curriculum describes the range of experiences which are necessary to render children capable of accepting the responsibilities which they inherit. It follows that children need to know and understand the characteristics of those materials and technologies which underpin our culture and language.

Some of our traditions and institutions have allowed us to make an erroneous separation between intellect and skill. We have thought, too often, of skill as a merely physical attribute and may have lost sight of the fact that we have the capacity to manufacture which informs – and most thoroughly challenges – our intelligence.

Conclusion

Design is a necessary accompaniment to technology and must draw upon all available knowledge to make informed propositions about its applications: these include wise use of material resources and accurate anticipation of consequences. The design profession's practice of simulating processes and outcomes by imaginative modelling, equates precisely with the activity by which children seek to develop capability. It is essential that children build up and retain confidence in their capacity to model and test their perceptions and ideas, and to compare them critically with others. They can then be expected to rise to meet the demands made by additional knowledge and by progressively more complex technology.

BIBLIOGRAPHY

Aldersey-Williams, H (1988) *New American Design.* New York: Rizzoli International Publications.

Angeloglou, M (1970) *A History of Make-up.* London: Studio Vista.

Arnheim, R (1967) *Towards a Psychology of Art.* London: Faber.

Arnheim, R (1970) *Visual Thinking.* London: Faber.

Asimov, I (1989) *Asimov's Chronology of Science and Discovery.* London: Grafton Books.

Atack, S M (1980) *Art Activities for the Handicapped.* London: Souvenir Press.

Atkinson, D (1991) 'How Children Use Drawing.' *Journal of Art and Design Education Vol.10, No.1.* Corsham: National Society for Education in Art and Design.

Attenborough, D (1987) *The First Eden: The Mediterranean and Man.* London: Collins/BBC.

Barnes, R (1987) *Teaching Art to Young Children 4-9.* London: Unwin Hyman.

Barnes, R (1989) *Art, Design and Topic Work 8-13.* London: Unwin Hyman.

Best, D (1985) *Feeling and Reason in the Arts.* London: George Allen & Unwin Ltd.

Blakemore, C (1976) *Mechanics of the Mind.* London: Cambridge University Press.

Bruner, J S (1972) *The Relevance of Education.* London: George Allen & Unwin Ltd.

Bruner, J S, Jolly, A, and Sylva, K (Eds) (1976) *Play: Its Role in Development and Evolution.* London: Penguin Books.

Chipp, H B (1968) *Theories of Modern Art.* Berkeley: University of California Press.

Cotterell, A (Ed) (1980) *The Encyclopedia of Ancient Civilizations.* London: Penguin.

Daumas, M (Ed) (1962) *A History of Technology and Invention.* Paris: Presses de Universitaires de France. Trans. (1969) New York: Crown Publishers.

de Chardin, P T (1959) *The Phenomenon of Man.* London: Collins.

Department of Education and Science and the Welsh Office. (1990) *Technology in the National Curriculum.* London: HMSO.

Design Council The, (1976) *Engineering Design Education.* (1977) *Industrial Design Education in the UK.* (1980) *Design Education at Secondary Level.* (1988) *Design in Primary Education.* London: The Design Council.

Dunn, S and Larson, R (1990) *Design Technology: Children's Engineering.* London: Falmer.

Eggleston, J (1976) *Developments in Design Education.* London: Open Books.

Everett, A (1986) *Materials.* London: The Mitchell Publishing Co.

Frayling, C (1987) *The Royal College of Art: One Hundred & Fifty Years of Art & Design.* London: Barrie & Jenkins.

Gardner, H (1978) *Developmental Psychology.* Boston: Little, Brown & Co.

Gardner, H (1983) *Frames of Mind.* London: Heinemann.

Garner, S W (1990) 'Drawing and Designing: the Case for Reappraisal.' *Journal of Art and Design Education Vol.9, No.1.* Corsham: National Society for Education in Art and Design.

Golomb, C (1974) *Young Children's Sculpture and Drawing.* New York: Harvard University Press.

Gombrich, E H (1960) *Art and Illusion.* London: Phaidon.

Gregory, R L (1981) *Mind in Science.* London: Weidenfeld & Nicholson.

Haviland, W A (1974) *Anthropology.* New York: Holt, Rinehart & Winston.

Haviland, W A (1975) *Cultural Anthropology.* New York: Holt, Rinehart & Winston.

Heskett, J (1980) *Industrial Design.* London: Thames & Hudson.

HM Inspectors of Schools (1978) *Primary Education in England.* London: HMSO.

Bibliography

Jelinek, J (1975) *The Evolution of Man*. Prague: Artia. Trans. by Hanks, H. London: Hamlyn.

Langer, S (1942) *Philosophy in a New Key*. New York: Harvard University Press.

Langer, W (Compiler/Ed) (1940 renewed 1968) *Encyclopaedia of World History*. London: Harrap.

Larsen, E (1960) *Ideas and Invention*. London: Spring Books.

Leakey, R (1982) *Human Origins*. London: Hamish Hamilton.

MacKinnon, J (1978) *The Ape Within Us*. New York: Holt, Rinehart & Winston.

Magee, B (1973) *Popper*. London: Collins.

Manzini, E (1989) *The Material of Invention*. London: The Design Council.

McGuinness, B F (1988) *Wittgenstein: A Life*. London: Penguin.

Meadows, J (1989) *Infotechnology: Changing the Way We Communicate*. Oxford: Equinox.

Medawar, P B (1972) *The Hope of Progress*. London: Methuen.

Moggridge, B (1991) 'The Craft Tradition.' *Journal No.1*. New York: American Centre for Design.

National Curriculum Council (1990) *Non-Statutory Guidance: Design and Technology Capability*. York: NCC.

Patten Woodhouse, C (1974) *The World's Master Potters*. London: David & Charles.

Popper, K R (1959) *The Logic of Scientific Discovery*. London: Hutchinson.

Premack, A J (1976) *Why Chimps Can Read*. New York: Harper & Row.

Pye, D (1978) *The Nature and Art of Workmanship*. London: Cambridge University Press.

Rawson, P (1969) *Drawing*. London: Oxford University Press.

Rawson, P (1983) *The Art of Drawing*. London: Macdonald & Co.

Read, H (1959) *A Concise History of Modern Painting*. London: Thames & Hudson.

Rowland, B (1965) *Cave to Renaissance*. Boston: Little, Brown & Co.

Rowling, M (1971) *Everday Life of Medieval Travellers*. London: Batsford.

Roy, R and Potter, S (1990) *Design and the Economy*. London: The Design Council.

Schools Council (1983) 'Primary Practice.' *Schools Council Working Paper 75*. London: Methuen.

Teissig, K and Brabcova, J (1982) *Drawing Techniques*. London: Octopus.

Thistlewood, D (Ed) (1990) *Issues in Design Education*. London: Longman/NSEAD.

Ward, C (1978) *The Child in the City*. London: The Architectural Press Ltd.

White, P (1974) *The Past is Human*. London: Angus & Robertson.

Wilkinson, G (1987) *A History of Britain's Trees*. London: Hutchinson.

Williamson, J (1978) *Decoding Advertisements: Ideology and Meaning in Advertising*. London: Boyars (Marion) Publishers Ltd.

Wilson, J and Rutherford (1989) 'Mental Models: Theory and Applications in Human Factors.' *Human Factors Journal, 31.6*. Santa Monica: Human Factors Society.

Wood, B (1978) *The Evolution of Early Man*. London: Peter Lowe.

Woods, K S (1975) *Rural Crafts of England*. London: Harrap & Co.

Young, J Z (1971) *An Introduction to the Study of Man*. London: Oxford University Press.

INDEX

adhesives 56, 63-5, 115, 118
  latex 119
  tape 118
agriculture, *technologies* 67
air 114-15, 119
aluminium 59
arrow heads 61
arrows 5
art and artists 107
  creative 120
  and design 70-4
  education 78, 82, 85-8
  experimental 72
  fine 87
  and materials 54
  and play 99
  in primary education 111
  response to materials 21-3
  teaching 86, 88-90
attainment targets, *design and technology* 8, 18, 45, 89-90, 124
  one 100-1
  two 101-3
  three 103
  four 103-4
axes 4, 61

behaviour, *symbolic* 32-3
belt drives 118
biotechnology 18
board, *compound* 63
bones 23, 55, 61-2, 67
  long 56
  tools 33
  as weapon 5
bows 5
brick 12, 62

card 63, 81, 114
  offcuts 118

sheet 119
casein (adhesive) 64
CDT (craft, design and technology) 84-5, 88-9
cement 36
chalk 2, 23, 55-6, 60
charcoal 23, 29-30, 55-6
children, *activity with materials* 80
  and confidence 8, 83, 96-7, 106, 131
  creative work 81
  as designers 127-9
  and experience 97-101
  experience for design 79
  learning 82-4
  and materials 104-5, 125-6
  modelling ideas 130-1
  and resource management 103
  respect for authority 8
  self-esteem 8
  and television 31
see also perception; play
classification 65-6
clay 56, 105, 118
  aboriginal material 55
  and access to materials 113
  containers 65
  'invented' materials 18
  limitations of 2-3
  as modelling material 30, 81, 119
  natural 63-5
  a natural plastic 63-4
  and play 94, 96
  slip 64-5
  writing tables 12
collage 119
colleges, *art* 85-6
  art and design 87, 91
  further and higher education 91

colour 107, 115-16
  and design 60, 72
  theory 116
communication 20, 44
  by drawing 82
  graphic 85
  media 28-9
  with machines 75
computer studies 90
computers 36, 63, 65-6, 76, 110
  communication with 126
  information 126
  intended users 126
concepts, 'good' and 'bad' 67
  number 61-2
Constructivism 70-1, 80, 87
containers 3, 56, 65-6, 118-19
  natural 65
cotton 15
crafts, *and design* 69-70
  education 78, 81
  design and technology (CDT) 84-5, 88-9
  guilds 38, 43
  and materials 21-3, 54
  in primary education 111
  regional 15
  schools 85-6
  traditions 16, 43
craftworkers 54, 107-8
  and access to materials 112-13
Cubism 72

darning 35
design, *awareness* 77, 79
  'Basic Design' movement 87
  and choice 17-18
  commercial 86
  for community life 66
  complex purposes 74-7

Council 90-1
craft-led 69
for crafts 87
critical 31
criteria 70, 77
decisions 75
education 41, 45
engineering 44
environment 49-53
evaluation 68, 103-4
exhibition and museum 87
failures 128
film set 87
fitness for purpose 70-1, 73-4
forward-looking 122
and freedom of mind 47-9
future-oriented 51-3
graphic 80, 82
human factors 74, 126
industrial 39, 44, 86
interior 87
and manufacture 43-5
National Diploma (1946) 86
and nature 70
necessity for 46-53
new factors 68-77
new possibilities 72-4
and new technology 18-19
and perception 49-50
and play 95, 97
and potential users 74-5, 97, 124
prejudice 104
principles of 85
processes 97
product 80, 87, 97
and quality 45
reactions to uniformity 71-2
related activity 79
responses to 68-9
and safety 75
School of (1837) 78

Index

schools of 85
specification 101-2
standards 45
strategies 44
and survival 32, 70, 74, 123
for survival 51
teaching strategies 125-6
technique development 41
three-dimensional 72, 87
and truth to material 69-70
uniformity in 71-2
and value judgements 76-7
see also design and technology capability; education; technology
design and technology, capability 46-7
education for 43, 87, 99, 123, 129
and materials 17-18, 23, 110, 131
new possibilities 89
reading and 27
diagrams, *flow* 116
domestic science 85
drawing 7, 23-4, 27, 29-30, 60-1, 106, 119
children's 30, 80-4
and design education 85-6, 103-4
engineering 84-5
and materials 129
and play 98
technical 84
drills 4
dung 3, 18, 55-6
bricks 62
dyes 56, 60-1, 115-16
colourfast 116

earth 55-6
bricks 62

substances 2
education, *access to materials* 111-13
authentic processes 107-8
background 78
business 90
for change and adaptability 123
child-centred 96
competition within 128
comprehensive 85
conceptual models 104-7
continuity and progress 129
cross-curriculum approach 89
design 78-91
for design and technology capability 43, 87, 99, 123, 129
design teaching 125, 129
further steps 122-9
as long-term investment 126-7
materials in 121
nursery and infant 115
and play 96-7
primary 80-4, 90-1, 100-9
secondary 79-80, 84-90
system 93-4, 127
vocational 78
What is it? 79-80
electronics 77, 85
energy sources 58
engineering 81-2, 95, 107
engraving, *fine tool* 105-6
wood 29
environment, *exploration* 101
hazards 112
and materials 112-13
modelling 103
examination, *A level* 86
system 39, 86
Expressionism 80-1

fabrics 115-16, 118
fashion 71-3, 87
fibre 55-6
animal 35
carbon 19
glass 3
natural 2-3, 36, 62, 118-19
synthetic 36, 118
vegetable 35
field studies 81
filament, *continuous* 118
film 30, 73
black and white 60
set design 87
finance 73
fishing 15
flight, *human-powered* 19
flints 4-5
hand-held 61
fluids 56, 58-60, 115, 119
food 13, 56, 67
design potential 120
dyes and pigments 116
plastic state 65
technology 16, 85
fuels 114
oil 59
resources 16

gases 55-6, 58-60, 115
glass 63-4, 87
fibre 3
glue 36
hoof-and-horn 64
PVA-based 119
goats, *effect on civilization* 12-13
graphics, *advertising and promotional* 87
computer 87
photographic 87
television 87
video 87
gravity 114

gum 60
natural 64
gut 3, 35, 55-6
'strings' 62

health and nutrition 67
home economics 85, 88, 90
hominids 2, 33
hunting 5

illustration, *book* 87
technical and scientific 87
illustrators 61
images 110
beyond 31-3
and changing perception 106
constructed 28-30
received 32
transient 111
video-recorded 30, 75, 110
virtual reality 76
visual 87-8
Impressionism 80
Industrial Revolution 15, 43, 59
industry 114
adaptation to changing nature of 16
art and design and 71-2
film 73
internationalization of 16
local 113-14
manufacturing 73
regional specialization 15-16
information, *from visual images* 111
new 19-20
storage and retrieval 65-6, 74
through senses 129
transfer of 20
information technology 7,

135

28, 75-6, 90, 124
   capability 89-90
ink 23, 60, 115
   printing 116
INSET *Video Training
   Packages* 108
instruments 56, 61-2,
   116-18
intelligence, *artificial* 75
   computer software 75
iron 64
   cast 59
   oxide 60
   ships 13
   and steel making in
     Sheffield 15
   wrought 59, 70

kinaesthetics 10-11, 57
knitting 62
knives 4
knowledge, *acquisition and
   development of* 40-1
   applied 38
   furtherance of 38-9
   institutional safeguards
     38-40
   training and standards
     37-45
   value of 37-8

language, *computer* 33
   and containers 65-6
   and design 76, 104-5
   development and
     materials 10-11
   education 88
   drawing and painting 61
   education in 19
   and formation of concepts
     129
   and large units 66
   and materials 13, 29,
     100-1, 110-11, 113, 115
   and modelling 117

and perception 26-7, 32
and plastics 65
and play 93
of research 20
and resources 50
and strings 62-3
technology and 13, 36,
   130
and tools 61-2
learning, *children's* 82-4
   continuing 3
   difficulties 129
   and drawing 82
   early 9-10
   from manufactured goods
     and toys 10
   kinaesthetic 10-11
   and materials 32, 115
   materials based 111
   and play 93-4, 99
   principles 115
   and survival 4, 18
   to be critical 30-1
   to classify 65
   to read 27-8
   to see 24-8, 31, 61, 104,
     124
   to use tools 10-11
lighting 41-2
local studies 81

make-up 60
manufacture, *large-volume
   production* 44
marketing, materials 121
market(s) 124
   international 73
   research 73-5
masks 60
materials, *aboriginal* 54-6,
   112-13
   absorbing new
     information 19-20
   access to 2, 101, 111-13
   alternative 105

in ancient and modern
   civilization 1-13
ancient civilization 11-13
animal and vegetable 67
available 26
behaviour 9, 22
building 119
categories 54-8, 67, 112,
   114-20
characteristics 6, 55, 69
for civilization 67
consequences of specialist
   employment 14-16
creative activity with 81
defined 1
dependence upon 12-14
and discipline 129
durable 33
exploitation 4
exploration and
   manipulation of 7, 94-5,
   130
found 6-7, 117, 130
hands-on experience 76,
   111
human responses to 21-3
and ideas 3, 54-67
influences 110-14
knowledge from 14
learning through activity
   with 7-11
limitations of 2-3, 30, 77
linear 35
live 94
local 113-14
location of 15
manufactured 35
marketing 121
modelling 30
natural 18, 113
nature of 22, 84
new 18, 20, 72, 77, 85, 94
new awareness 14-20
in perception 26-7
physical attributes and the

   mind 3-7
and play 92-9
potential 37, 46
practical activity with
   127-9
principles of movement
   and structure 10
properties 17-18, 21-3,
   27, 29, 55, 100, 121
qualities 69, 104
raw 113
ready-made 114
resources 6, 15, 17, 122,
   131
resources, living and non-
   living 18
respecting 50
rigid and semi-rigid 36
safety and danger 120-1
salvaged 117
structural limitations 2-3
and survival 31-2, 43
synthetic 19
technological development
   17-19
technology 85
three-dimensional 98
traditional disciplines
   21-36
truth to, in design 69-70
uses made of 5
valuing limitations of 2-3
world 18
see also language;
   perception
mechanics 58
media, *communication* 28-9
   news, limitations of 28-9
   pictorial 29
   technology 31
metals 18, 23
   sheet 119
   smelting 59
metalwork 84
minerals 61

Index

mining 15
models and modelling 3, 7, 26-7, 50-1, 82, 106, 129
　for abstract ideas 33
　additive 119
　children's designs 128
　clockwork 107
　conceptual 59, 116
　for construction of ideas 19, 22
　and design experience 101, 103-4
　and design teaching 126
　designing and assembling 116
　and development of civilization 12
　of ideas 83, 130
　imaginative 131
　'junk' 81, 117
　materials for 30
　mental 112, 117
　'mock-up' 128
　perceptual 25
　plastic 65
　and play 98
　three-dimensional 114, 119
　of time, space and energy 66
　two-dimensional 114
　wheeled 114, 128
　working 81
monoliths 66

National Curriculum 18, 40, 99, 122-3, 125
　and access to materials 111
　art in 127
　Council 90
　Council for Wales 90
　design 124
　design and technology 89, 127, 129-30
　design education in 79
　technology 87
　see also attainment targets
needles 61-2
newspapers 28
number concepts 61-2

oak, *for ships* 13
objects 107-8
　'found' 107
oils 59-60

packing and packaging 63, 87, 107
paint and painting 80, 106, 118-19
　block 115
　landscape 105
　liquid 115
　and materials 129
　opaque 60, 115
　and play 96, 98
　powder 115
　translucent 60
　transparent 115
paper 63, 81, 114
　sheets 118
　smooth and textured 23
papyrus, *technology* 35
pencils 23, 98
perception, *changing* 105-7
　children's 37, 49, 125-6, 128, 131
　and design 49-50, 68
　and language 32
　and materials 26-7, 31-3
　and new design 72
　with and without words 25-6
photocopier 121
photography 28-30, 106
　graphics 87
　still 30
picks 61
pictures 26-7, 106

mental 28, 60-1
pigments 56, 60-1, 115-16
pigs, *and ships* 13
plaster 119
plastics 56, 63-5, 114
　modern 64
　opaque 63
　properties 63
　sheet 119
　tape 118
　thermosetting 64
　translucent 63
　transparent 63
play 62, 92-9, 123-4
　criteria 96
　and design 101-3
　instinct to 92-4
　pleasure from materials 94-5
　and reality 95-7, 102
　role-play 96-7
　rules of 95, 97
　supporting imaginative 95-6
　trial-play 97
pliers 56
poison 67
polyester 19
polytechnics **86, 91**
pottery 3, 65
　Staffordshire 15, 69
pre-fabrication 113
printer 121
printing 35, 60, 87
　block 115
　on fabrics 115-16
　inks 116
　photographic 119
　screen 115
　technology 110
PVA (adhesive) 64, 115, 119

quality and design 45

reeds 3, 35, 55-6
research, *local* 108
　market 73-5
resin 64
resources, *responsibility for* 20
　see also materials
rods and levers 56-8, 96, 114-15
role-play 96-7
rope, *invention of* 34-5

scale 66
schools, *sixth-form courses* 86-7
　see also education
science, *and access to materials* 111
　and design 74
　physical 113
　and play 99
　response to materials 21-3, 54
scissors 56
scrapers 4, 33, 61
　flint 5
sculpture 66, 72
　three-dimensional 81
'self-expression' 81
senses 25-7, 60, 68, 130
sewing 35
sheet(s) 56, 63-5, 119
　containers 65
　flexible 35-6
　manufactured 35
　metal 63
　paper 118
　synthetic 63
　technology 33-6
shipbuilding 12-13, 15
size (adhesive) 64
skin 23, 55-6, 63
　fur-bearing 34
　and rope invention 35
　and sheet technology 33-4

137

socialization 5, 93-4
solder 64-5
solvents 64
sound and volume 119
space, well-designed 66
spear heads 4
spears, *fire-hardened* 5
specialization 39
  spinning and weaving 15, 34-5
standardization 44
standards 42, 44, 73-4
  in design 45, 76
  and design education 85-6
  institutional safeguards 38-40
  knowledge and training 37-45
steel 15, 59
stone 2-3, 23, 33, 55, 62, 66
  and ancient civilization 11-12
  and play 94
  as tools and weapons 4-5
strings 35, 56, 62-3, 118-19
sugars 64
surfaces 107
  flat and relief 119
symbolization 61-2

tattooing 60
teachers, *and access to materials* 111
  art 105
  art and design 86, 88-90
  and attainment targets 101-4
  business education 90
  CDT 89-90
  computer studies 90
  confidence 109
  design 76-7, 85
  as designers 108-9, 126-7
  education 122-3, 130
  further and higher education 86, 105, 108
  head 87
  home economics 89-90
  how to use 126
  information technology 90
  and marketing 121
  and materials 1, 67, 114
  and models 117-18
  and play 94-5, 97-9
  primary-school 37, 80-2
  safety and danger 120
  skills and intellect 40
  specialist 108
  support for activity 19
  supportive intervention 7
  training of 126-7
  witnessing learning 7
  of young children 9
technology, *and access to materials* 111
  and changing perception 106
  and choice 17-18, 46-7
  complex 131
  defined 1
  and design 46-7, 108-9, 131
  development 2, 5
  and economy 130
  evaluation 104
  evolution 122-4
  ideas and language 130
  importance of 6
  information 28
  innovation 33
  and location of industry 15
  and materials 54
  media 31
  and need for design 46-7
  new developments 18-19, 40
  origins and nature of 2, 6
  of papyrus 35
  in primary education 40
  problems 114
  and resources for education 127
  sheet 33-6
  of shipbuilding 12-13
  space 17
  and survival 46
  television 17
  textile 35-6
  and thought 41-2
  transport 16
  users of 67
  and value judgements 76-7
television 17, 28-9, 31, 76
  colour 30
  graphics 87
textiles 18, 23, 63, 119
  and design 72, 87
  education 85
  technology 35-6, 62
  waste 118
  woven 35
texture 107
thermoplastic 64
tools 1-2, 61-2, 116-18
  automation 113
  basic 116, 118
  bone 33
  early 4-5
  fine 105-6
  first 33-4
  flint 4
  'found' 56
  and learning 10-11
  and learning development 4
  limitations of 31
  multi-purpose 74
  origins of 3
  'ready-made' 56
  single-purpose 73-4
  technology and 6, 47-8
  use 52-3
training, *development of* 38-9
  standards and knowledge 37-45
transport, **technology** 16
trees 55-6
  drawing 22
  human affinity with 66
  typography 87

uniformity in design 71-2
units, *large* 56, 66, 120
  small 56, 61-2, 116-18
universities 86

video 30, 36, 75, 110
virtual reality 76
visits 108

water 55-6, 59-60, 114-15
  as energy source 58
  and play 96
weapons **4-5, 42**
weaving 15, 34-5, 62
welding 64
wind, *as energy source* 58
wood 23, 33, 81, 118
  in design 70
  engravings 29
  and play 96
  for shipbuilding 12-13
woodwork 84
wool 15, 113
word processor 121

yarns 107, 118